STUDENT SOLUTIONS MANUAL

TO ACCOMPANY

CALCULUS
FROM GRAPHICAL, NUMERICAL, AND SYMBOLIC POINTS OF VIEW

SECOND EDITION   VOLUME 2

OSTEBEE/ZORN

Arnold Ostebee
St. Olaf College

HOUGHTON MIFFLIN COMPANY   BOSTON NEW YORK

Sponsoring Editor: Lauren Schultz
Assistant Editor: Marika Hoe
Senior Manufacturing Coordinator: Florence Cadran
Senior Marketing Manager: Michael Busnach

Printed in the U.S.A.

ISBN: 0-618-25413-7

1 2 3 4 5 6 7 8 9 – PAT – 06 05 04 03 02

TABLE OF CONTENTS

M MULTIVARIABLE CALCULUS: A FIRST LOOK

6   *Numerical Integration*

§6.1   Approximating Integrals Numerically

1. (a) $I = \int_1^4 \dfrac{dx}{\sqrt{x}} = 2\sqrt{x}\,\Big]_1^4 = 4 - 2 = 2.$

    (b) $L_3 = \dfrac{4-1}{3}\left(f(1) + f(2) + f(3)\right) = \dfrac{1}{\sqrt{1}} + \dfrac{1}{\sqrt{2}} + \dfrac{1}{\sqrt{3}} \approx 2.28446;$

    $R_3 = \dfrac{4-1}{3}\left(f(2) + f(3) + f(4)\right) = \dfrac{1}{\sqrt{2}} + \dfrac{1}{\sqrt{3}} + \dfrac{1}{\sqrt{4}} \approx 1.78446;$

    $T_3 = (L_3 + R_3)\,/2 \approx 2.03446;$

    $M_3 = f\left(\dfrac{3}{2}\right) + f\left(\dfrac{5}{2}\right) + f\left(\dfrac{7}{2}\right) \approx 1.98347.$

    (c) $|I - L_3| \approx 0.28446;\ |I - R_3| \approx 0.21554;$
    $|I - T_3| \approx 0.03446;\ |I - M_3| \approx 0.01653.$

    (d) $S_6 = \dfrac{2}{3}M_3 + \dfrac{1}{3}T_3$

    $\approx \dfrac{2}{3}(1.98347) + \dfrac{1}{3}(2.03446)$

    $\approx 2.00047.$

    Therefore, the approximate error using S_6 is 0.00047.

3. The integrand $f(x) = \sin(x^2)$ is monotone increasing on [0, 1], so all left sums underestimate and all right sums overestimate the integral.

5. (a) $L_{10} = 11.810;\ R_{10} = 22.530;\ T_{10} = 17.170;\ M_{10} = 16.098;\ S_{20} = 16.455.$

    (b) $|I - T_{10}| \leq |L_{10} - R_{10}| = 10.72.$

    (c) $|I - S_{20}| \leq |T_{10} - M_{10}| = 1.072.$

7. (a) We'll do L_4 explicitly; the others are similar.

    $$L_4 = \dfrac{1}{4}\left(f(0) + f(0.25) + f(0.5) + f(0.75)\right) = 1.1485.$$

    Similar calculations show:

    $$R_4 = 1.1805; \quad M_2 = 1.1345; \quad T_4 = 1.1645; \quad S_4 = 1.1545.$$

    (b) Upward concavity means that T_4 must *overestimate* and M_2 must *underestimate*. Therefore I has to lie in the interval [1.1345, 1.1645]. (The S_4 estimate is consistent with this.)

---

9. (a) Calculating I by antidifferentiation gives $I = 0.45970$ (to 5 decimals). Combining this with the tabulated data gives $I - L_{20} \approx 0.02113$; $I - R_{20} \approx -0.02094$; $I - T_{20} \approx 0.00010$; $I - M_{20} \approx -0.00005$.

    (b) L_{20} underestimates I because the integrand is increasing on $[0, 1]$.

    (c) T_{20} underestimates I because the integrand is concave down on $[0, 1]$.

11. All entries are negative; this means that $M_n > I$ in each case.

13. $S_{200} = (2M_{100} + T_{100})/3 \approx 0.31027$; note that M_{100} and T_{100} are given in Example 3.

15. Any decreasing function will do; $f(x) = 1/x$ is one possibility.

17. Any function that's concave down on $[1, 5]$ will do; $f(x) = 1 - x^2$ is one possibility.

19. (a) Upward concavity means that the graph "sags beneath" the straight line segments that define the trapezoid rule.

    (b) Yes: T_n still overestimates I for all $n \geq 1$. Concavity matters, but not positivity or negativity of f.

21. By definition, $L_n = \big(f(x_0) + f(x_1) + \cdots + f(x_{n-1})\big) \Delta x$ and $R_n = \big(f(x_1) + f(x_2) + \cdots + f(x_n)\big) \Delta x$. Adding these formulas and dividing by two gives $T_n = \left(\dfrac{f(x_0)}{2} + f(x_1) + \cdots + f(x_{n-1}) + \dfrac{f(x_n)}{2} \right) \Delta x$, as desired.

23. For *any* function, S_{2n} lies between M_n and T_n. This interval has length $|M_n - T_n|$. If f never changes direction of concavity, then I must lie in this same interval. Thus both S_{2n} and I lie in the same interval, and so can't differ by more than $|M_n - T_n|$, the interval's length.

25. A picture similar to that in the discussion of Theorem 2 makes this clear.

27. Let $I = \displaystyle\int_1^7 f(x)\,dx$. Since f is positive on the interval of integration, all five estimates are positive numbers. Since f is increasing on the interval of integration, $L_{100} \leq I \leq R_{100}$. Since f is concave down on the interval of integration, $T_{100} \leq I \leq M_{100}$. Furthermore, since T_{100} is the average of L_{100} and R_{100}, $L_{100} \leq T_{100} \leq R_{100}$. Also, S_{200} is a weighted average of T_{100} and M_{100}, so $T_{100} \leq S_{200} \leq M_{100}$. Therefore,

$$L_{100} \leq T_{100} \leq S_{200} \leq M_{100} \leq R_{100}.$$

29. Must: Left sums underestimate.

31. Must: On each subinterval f is larger at the right endpoint than at the left endpoint.

33. Must: We know $L_n < R_n$, and T_n is the average of L_n and R_n.

35. May: It depends on whether f is increasing or decreasing.

37. May: It depends on whether f is increasing or decreasing.

39. May: It depends on whether f is increasing or decreasing.

41. Must: f decreasing means that all right sums underestimate.

43. Cannot: f decreasing means that left sums overestimate and right sums underestimate.

45. Must: On each subinterval f is larger at the midpoint than at the right endpoint.

47. Must: Because $f' > 0$, we know f is increasing, so left sums underestimate.

49. **Must** be true. Since f is increasing on the interval of integration $M_n < R_n$ for every n. [If f is increasing on an interval $[a, b]$, then $m = (a + b)/2$ is the midpoint of the interval and $f(a) < f(m) < f(b)$.]

51. The integrand's concavity changes twice on the interval $[0, 2]$, once near $x = 0.808$ and again near $x = 1.814$. Breaking up the integral at these points allows the sort of trapping desired.

53. (a) Since f is linear, the trapezoid approximation "fits" f exactly over each subinterval.

   (b) A picture shows that the trapezoidal area is equal to a rectangular area.

   (c) The preceding parts show that, for a linear function, the midpoint and trapezoid estimates are identical.

55. Recall that $T_n = (L_n + R_n)/2$. Into this, substitute the expression for R_n derived in the previous exercise.

57. Suppose first that $n = 1$. Because f is increasing, the f-graph lies inside the rectangle with lower left corner $(a, f(a))$ and upper right corner $(b, f(b))$. (Draw a picture!) Because f is concave up, the f-graph sags below the diagonal line from $(a, f(a))$. A close look at the picture shows that the midpoint estimate M_1 approximates I better than both L_1 and R_1. The same argument (applied to each subinterval) works if $n > 1$.

59. The information given implies that f is increasing and concave up on the interval $[a, b]$. A sketch shows that the approximation error made by R_n includes all of the area corresponding to the approximation error made by T_n and more. Thus the answer is "no."

61. Since f' is negative on $[a, b]$, f is decreasing; therefore, $R_n \le I \le L_n$, $M_n \le L_n$, and $R_n \le T_n$. Since f' is decreasing on $[a, b]$, f is concave down; therefore, $T_n \le I \le M_n$. Putting these pieces together, we have $R_n \le T_n \le I \le M_n \le L_n$.

63. The picture shows that $\left(f'(x)\right)^2$ is increasing; thus, so is the integrand. This implies that R_n overestimates the integral.

65. The graph shows $f'(x)$; since $f' < 0$ and f' is increasing, the function f itself must be decreasing and concave upward. It follows that $R_n \le M_n \le S_{2n} \le T_n \le L_n$. (Note that S_{2n} lies between M_n and T_n for every integral.)

67. Let $g(x) = \left(f(x)\right)^2$; then $g'(x) = 2f(x)f'(x)$. Since $f(x) < 0$ and $f'(x) < 0$ for $x > 0$, we have $g'(x) > 0$, so g is increasing. Thus, $L_n \leq I \leq R_n$.

69. The graph shows $f'' > 0$, so f is concave up. Thus M_n must underestimate I.

71. The graph shows that f'' is positive—hence f is *concave up*—over the interval of integration.

    $L_{100} \leq R_{100}$ could be true or false, depending on whether f is increasing or decreasing. The graph of f'' doesn't tell which is the case.

73. $M_{50} \leq L_{50}$ could be true or false, depending on whether f is increasing or decreasing. The graph of f'' doesn't tell which is the case.

§6.2   Error Bounds for Approximating Sums

1. (a) Plotting f'' over $[0, 3]$ shows that $K_2 = 32$ is suitable. Plugging $K_2 = 32$, $a = 0$, $b = 3$, and $n = 50$ into the error bound formula gives $|I - M_{50}| \leq 0.0144$, which is well below 0.02.

   (b) Here $M_{50} \approx 0.774$; part (a) implies that I lies within 0.02 of this number. That is, $0.754 \leq I \leq 0.794$.

3. All three integrands are linear functions, so their derivatives are constants. The values of $|f'|$ are, respectively, 1, 2, and 4; the absolute errors committed by L_4 (0.125, 0.250, and 0.500) are in the same proportion.

5. (a) Using $K_1 = 5$, $a = 0$, $b = 2$, and $n = 8$ in Theorem 3 gives $|I - R_8| \leq 5/4$.

   (b) Using $K_2 = 0$ in Theorem 3 gives $|I - T_n| = 0$ for all n.

7. (a) Plotting f' shows $|f'(x)| \leq 1.3$ on $[0, 1]$; thus any $K_1 \geq 1.3$ is acceptable.

   (b) Plotting f'' shows $|f''(x)| > 2$ near $x = 1$, so 2 is not an acceptable value for K_2.

9. (a) $I = \int_1^4 \dfrac{dx}{\sqrt{x}} = 2\sqrt{x}\,\Big]_1^4 = 4 - 2 = 2$.

   (b) Calculation gives $L_3 \approx 2.28446$ and $R_3 \approx 1.78446$. Comparing these with the exact answer, 2, gives $|I - L_3| \approx 0.28446$ and $|I - R_3| \approx 0.21554$. The error bound from Theorem 3 is found by using $K_1 = 0.5$, $a = 1$, $b = 4$, and $n = 3$; the result is 0.75, which is well above the actual errors committed.

   (c) Calculation gives $T_3 \approx 2.03446$ and $M_3 \approx 1.98347$. Comparing these with the exact answer, 2, gives approximation errors $|I - T_3| \approx 0.03446$ and $|I - M_3| \approx 0.01653$. The error bounds from Theorem 3 are found by using $K_2 = 0.75$, $a = 1$, $b = 4$, and $n = 3$. This gives $|I - T_3| \approx 0.18750$ and $|I - M_3| \approx 0.09375$, which are well above the actual errors committed.

   (d) With $K_2 = 0.75$, $a = 1$, and $b = 4$, Theorem 3 says that $|I - M_n| \leq \dfrac{0.84375}{n^2}$. The last quantity is less than 0.005 if $n \geq 13$; this guarantees that M_n has two decimal place accuracy if $n \geq 13$.

11. For $\int_0^3 e^{-x^2}\, dx$ we have $f'(x) = -\exp(-x^2)2x$. On $[0, 3]$, $|f'(x)| \leq 0.8578$, so $K_1 = 0.8578$ works. Hence we need $n \geq 0.8578 \cdot 3^2 \cdot 100 = 772.02$ so $n \geq 772$. (If K_1 is approximated, slightly different results occur.)

13. For $\int_0^1 \left(1 + x^2\right)^{-1} dx$, $K_1 = 0.65$ works so any $n \geq 0.65 \cdot 1^2 \cdot 100 = 65$ will do.

15. For $\int_0^3 e^{-x^2}\, dx$, $K_2 = 2$ works. Setting $K_2 = 2$, $a = 0$, and $b = 3$ in Theorem 3 shows that we want n so $\dfrac{2 \cdot 3^3}{24n^2} < 0.005$. Solving for n gives $n \geq 22$.

17. For $\int_0^1 (1 + x^2)^{-1}\, dx$, $K_2 = 2$ works. Setting $K_2 = 2$, $a = 0$, and $b = 1$ in Theorem 3 shows that we want n so $\dfrac{2 \cdot 1^3}{24n^2} < 0.005$. Solving for n gives $n \geq 5$.

19. Any linear and increasing function will work; $f(x) = x$ is one possibility.

21. If $f(x) = x^2$, then f is concave up, so M_n underestimates I and the approximation error is as bad as Theorem 3 allows.

23. (a) Doubling n halves the error bound.

    (b) For M_n and M_{2n}, doubling n reduces the error bound by a factor of 4.

    (c) Parts (a) and (b) suggest that, as a rule, the midpoint and trapezoid rules perform better than the left and right rules.

25. With no bound given for K_1 we have no accuracy guarantees.

27. (a) For $\int_1^{11} \cos(1/x)\, dx$, $K_1 = 0.8415$ works. Setting $K_1 = 0.8415$, $a = 1$, and $b = 11$ in Theorem 3 shows that we want n so $\dfrac{0.8415 \cdot 10^2}{2n} < 0.005$. Solving for n gives $n \geq 8415$.

    (b) For $\int_1^6 \cos(1/x)\, dx$, $K_1 = 0.8415$ works. Setting $K_1 = 0.8415$, $a = 1$, and $b = 6$ in Theorem 3 shows that we want n so $\dfrac{0.8415 \cdot 5^2}{2n} < 0.004$. Solving for n gives $n \geq 2630$.

    (c) For $\int_6^{11} \cos(1/x)\, dx$, $K_1 = 0.005$ works. Setting $K_1 = 0.005$, $a = 6$, and $b = 11$ in Theorem 3 shows that we want n so $\dfrac{0.005 \cdot 5^2}{2n} < 0.001$. Solving for n gives $n \geq 63$.

    (d) Adding estimates for the two partial integrals gives an estimate for the total integral; the total error is guaranteed less than $0.004 + 0.001 = 0.005$.

    (e) Using the results above gives a good estimate with about 2700 values of f—a lot less than the original 8415 values of f.

29. (a) $I = \pi$ because the integral gives the area of the "northeast quadrant" of a circle of radius 2.

    (b) $L_{10} \approx 3.3045$; $|I - L_{10}| \approx 0.1629$.

    (c) Theorem 3 doesn't give a good bound here because we can't compute K_1 — $f'(x)$ is unbounded on the interval $(0, 2)$.

31. Since $F'' = f'$, the graph shows that we can take $K_2 = 9$. Setting $K_2 = 9$, $a = 1$, and $b = 2$ in Theorem 3 shows we want n so $\dfrac{9 \cdot 1^3}{24n^2} < 0.01$. Solving for n gives $n \geq 7$.

33. Since $\left| f''(x) \right| \leq 2.5$ if $0 \leq x \leq 5$, $\left| T_n - \int_a^b f(x)\, dx \right| \leq 2.5 \cdot 5^3 / 12n^2$. The expression on the right is less than 0.001 if $n \geq 162$.

35. Cannot: To estimate the error committed by L_{100}, we use $K_1 = 5$, so $|I - L_{100}| \le 5 \cdot 10^2/200 = 2.5$. Since $I = 9$, we must have $L_{100} < 11.5$.

37. Cannot: Because h is concave up, T_n must overestimate $I = 9$.

39. For a monotone function, I lies *between* L_n and R_n. The distance between L_n and R_n is $|R_n - L_n|$, and a little algebra shows that this difference is $|f(b) - f(a)| \cdot (b-a)/n$.

41. For a monotone function, the "exact" integral I lies *between* L_n and R_n, i.e., in an interval of length $|R_n - L_n| = |f(b) - f(a)| \cdot (b-a)/n$. Since T_n is the *midpoint* of this interval, I must lie within *half* the interval's width from T_n. (Draw L_n, R_n, and T_n on a number line to understand all this.)

§6.3 Euler's Method: Solving DEs Numerically

1. $y(3) \approx y(2) + y'(2) \cdot 1 = 167.20 - 9.72 = 157.48$.

3. Differentiation shows that if $y(t) = 70 + 120e^{-0.1t}$, then $y'(t) = -12e^{-0.1t} = -0.1(y - 70)$, as desired.

5. (a) The DE $P' = kP(C - P)$ implies that $P'' = kP'(C - P) - kPP' = kP'(C - 2P)$; this is negative when $2P > C$, or $P > C/2$. That's the case here because $P(0) = 6000 > C/2 = 5000$.

   (b) Since $P(t)$ is concave down and Euler's method is based on tangent lines, Euler's method will overestimate, since tangent lines lie above the graph of $P(t)$.

7. (a) $f\big(t_i, Y(t_i)\big) = m_i$.

   (b) $f\big(t_{i+1}, Y(t_{i+1})\big) = m_{i+1}$.

9. (a) With just 1 subdivision, Euler's method gives

$$e = y(1) \approx y(0) + y'(0) \cdot 1 = 1 + 1 \cdot 1 = 2.$$

   This answer *underestimates* e. We can tell this even without knowing the true value of e. The DE $y' = y$ means that $y'(t)$ increases over the interval $0 \leq t \leq 1$; taking just one Euler step over the interval amounts to pretending that y' is *constant* over this interval.

   (b) Carrying out Euler's method with 4 subdivisions, from $t = 0$ to $t = 1$, gives the following table:

step	t	y'	y
0	0.00	1.0000	1.0000
1	0.25	1.2500	1.2500
2	0.50	1.5625	1.5625
3	0.75	1.9531	1.9531
4	1.00	2.4414	2.4414

   In particular, $e = y(1) \approx 2.4414$. For the same reason as in the previous part of this exercise, this result *underestimates* the true value of e.

   (c) In this case Euler's method underestimate the exact solution for *every* n.

   (d) The first few steps show the pattern for $n = 1, 2, 3$:

$$y(1/n) = 1 + 1 \cdot \frac{1}{n}$$

$$y(2/n) = \left(1 + \frac{1}{n}\right) + \left(1 + \frac{1}{n}\right) \cdot \frac{1}{n} = \left(1 + \frac{1}{n}\right)^2$$

$$y(3/n) = \left(1 + \frac{1}{n}\right)^2 + \left(1 + \frac{1}{n}\right)^2 \cdot \frac{1}{n} = \left(1 + \frac{1}{n}\right)^3.$$

   The same pattern holds for all n.

11. (a) Results can be tabulated as follows:

t	0	0.4	0.8	1.2	1.6	2.0
y'	0	0.4	0.64	0.784	0.870	0.922
y	1	1	1.16	1.416	1.730	2.078

They show that the Euler estimate is $Y(2) = 2.078$.

(b) Calculating as in (a), but with step size 0.2, gives the Euler estimate is $Y(2) = 2.107$.

(c) If $y = t + e^{-t}$, then a direct check shows that (i) $y' = 1 - e^{-t} = 1 + t - y$, and (ii) $y(0) = 1$, as required.

(d) The exact value of $y(2)$ is $y(2) = 2 + e^{-2} \approx 2.135$—a little more than both Euler estimates. This underestimate could have been predicted from the slope field, which shows that solutions are concave upward.

13. (a) $y(1) \approx -0.59374$

(b) $y(1) \approx -0.65330$

(c) $y'(t) = -e^t = (2 - e^t) - 2 = y(t) - 2$ and $y(0) = 1$

(d) $y(1) = 2 - e \approx -0.71828$.

15. (a) $y(0.8) \approx 2.6764$

(b) $y'(t) = (1 - t)^{-2} = \big(y(t)\big)^2$ and $y(0) = 1$

(c) The derivative is changing rapidly and is becoming large. Thus, very small steps are required to achieve an accurate result.

17. (a) If $y = 1 - e^{-t}$, then a direct check shows that (i) $y'(t) = e^{-t}$ and (ii) $y(0) = 0$, as desired.

(b) Results can be tabulated as follows:

t	0	0.2	0.4	0.6	0.8	1.0
Y'	1	0.819	0.670	0.549	0.449	0.368
Y	0	0.2	0.364	0.498	0.608	0.697

They show that the Euler estimate is $Y(1) \approx 0.697$.

(c) Is a plot.

(d) Since y' is a decreasing function, any left sum estimate of $\displaystyle\int_0^T y'(t)\,dt$ will overestimate the exact answer.

7 *Using the Integral*

§7.1 Measurement and the Definite Integral; Arclength

1. (b) The area is approximately $L_5 = \dfrac{1}{5} \sum\limits_{i=0}^{4} \left(\dfrac{i}{5} - \dfrac{i^2}{25} \right) = \dfrac{4}{25} = 0.16$.

   (c) The left sum approximation of the area under the curve $y = x - x^2$ *underestimates* the actual area under the curve.

   (d) The area is $\displaystyle\int_0^1 (x - x^2)\, dx = \dfrac{1}{2}x^2 - \dfrac{1}{3}x^3 \Big]_0^1 = \dfrac{1}{6}$.

3. (a) Note $x = 1 - y^2$ implies $y = \sqrt{1 - x}$, so the area is $\displaystyle\int_0^1 \left(1 - \sqrt{1 - x} \right) dx = 1/3$.

   (b) The region is symmetric with that of (a), so its area is also $1/3$.

   (c) Note $x = y + 2$ implies $y = x - 2$, so the area is
   $$\int_1^3 \big(1 - (x - 2) \big)\, dx = \int_1^3 (3 - x)\, dx = 3x - \dfrac{1}{2}x^2 \Big]_1^3 = 2.$$

   (d) (a) + (b) + (c) = $8/3$.

5. (a) Applying T_{20} to the given integral I gives $T_{20} \approx 1.47932$.

   (b) Because the integrand is concave up, all midpoint sums underestimate and trapezoid sums overestimate.

7. Because $dy/dx = \cos x$, the length integral is $I = \displaystyle\int_0^\pi \sqrt{1 + \cos^2 x}\, dx$. For this integral, $M_{20} \approx 3.8202$.

9. The integral has the same *numerical* value in each part:
   $$\int_0^3 f(t)\, dt = \int_0^3 \left(5t^2 - 20t + 50 \right) dt = 105. \text{ The parts differ from each other only in}$$
   interpretation.

   (a) The car traveled 105 miles during the three hours from midnight to 3 am.

   (b) The car traveled 105 feet in the first 3 seconds after midnight.

   (c) The car's velocity increased by 105 feet per minute over the first 3 minutes after midnight.

11. (a) Estimates will vary depending on which difference quotients are used. The following estimates are computed using "symmetric" difference quotients—e.g.,

$f'(1.0) \approx (f(1.1) - f(0.9))/0.2$:

t	1.0	1.1	1.2	1.3	1.4	1.5
$f'(t)$	-0.83	-1.17	-1.46	-1.70	-1.875	-1.965
t	1.6	1.7	1.8	1.9	2.0	
$f'(t)$	-1.985	-1.925	-1.78	-1.57	-1.3	

(b) Different estimates are possible using the data from (a). For example, $M_5 = -1.666$. Here $f(2) - f(1) = -1.666$, too.

(c) Different numerical integral estimates are possible using the data from the table. For example, $M_5 = 0.1194$. This number estimates the signed area determined by the f-graph.

(d) Data from the table in (a) can be used to estimate the integral; in this case $M_5 = 1.9494$. This integral gives the length of the f-graph from $t = 1$ to $t = 2$.

(e) The integrals in question are $I_1 = \int_1^2 2\cos(2t)\,dt$ and $I_2 = \int_1^2 \sin(2t)\,dt$. Evaluating them exactly gives $I_2 = \cos(2)/2 - \cos(4)/2 \approx 0.1187$ and $I_1 = \sin(4) - \sin(2) \approx -1.6661$.

13. Area $= \displaystyle\int_{-1}^{1} (1 - x^4)\,dx = \frac{8}{5}$.

15. Area $= \displaystyle\int_{0}^{1} (x^2 - x^3)\,dx = \frac{1}{12}$.

17. Area $= \displaystyle\int_{0}^{4} \sqrt{x}\,dx = \frac{16}{3}$.

19. Area $= \displaystyle\int_{-\frac{1}{4}}^{1} \left[(2 - x) - \frac{9}{4x + 5} \right] dx = \frac{65}{32} - \frac{9}{4}\ln(9) + \frac{9}{4}\ln(4) \approx 0.20666$.

21. Area $= \displaystyle\int_{0}^{1} e^x\,dx = e - 1$.

23. The curves intersect at the points $(0, 2)$ and $(5, -3)$. The area is found in two parts, first from $x = -4$ to $x = 0$ and then from $x = 0$ to $x = 5$. The result is
$$\text{Area} = 2\int_{-4}^{0} \sqrt{4 + x}\,dx + \int_{0}^{5} (2 - x + \sqrt{4 + x})\,dx = \frac{32}{3} + \frac{61}{6} = \frac{125}{6}.$$

25. Note that $y = x^4$ implies $x = \pm y^{1/4}$. Thus Area $= 2\displaystyle\int_{0}^{1} y^{1/4}\,dy = \frac{8}{5}$.

27. Note that $y = x^3 \implies x = \sqrt{y}$ and $y = x^3 \implies x = y^{1/3}$. Thus,
$$\text{Area} = \int_{0}^{1} (y^{1/3} - y^{1/2})\,dy = \frac{1}{12}.$$

29. Note that $y = \sqrt{x} \implies x = y^2$. Thus, Area $= \int_0^2 (4 - y^2)\,dy = \dfrac{16}{3}$.

31. The curves have equations $x = \dfrac{9}{4y} - \dfrac{5}{4}$ and $x = 2 - y$; they cross at the points $(1, 1)$ and $(-1/4, 9/4)$. Thus, Area $= \int_1^{9/4} \left(2 - y - \dfrac{9}{4y} + \dfrac{5}{4}\right) dy = \dfrac{65}{32} - \dfrac{9}{4} \ln \dfrac{9}{4} \approx 0.2067$.

33. The bounding "curves" have equations $x = \ln y$ and $x = 1$. Thus,
$$\text{Area} = \int_0^1 (1 - 0)\,dy + \int_1^e (1 - \ln y)\,dy = 1 + (e - 2) = e - 1.$$

35. The curves intersect at $y = -3$ and $y = 2$, so
$$\text{Area} = \int_{-3}^2 [(2 - y) - (y^2 - 4)]\,dy$$
$$= \int_{-4}^0 2\sqrt{x + 4}\,dx + \int_0^5 \left[(2 - x) + \sqrt{x + 4}\right] dx = \dfrac{125}{6}.$$

37. Let the pizza be defined by the circle $x^2 + y^2 = 36$. Then the middle "third" has area
$$4 \int_0^2 \sqrt{36 - x^2}\,dx \approx 47.096 \text{ square inches, while the other pieces have area}$$
$$2 \int_2^6 \sqrt{36 - x^2}\,dx \approx 33.001 \text{ square inches. The ratio of areas is } 47.096/33.001 \approx 1.4271.$$

39. length $= \int_a^b \sqrt{1 + f'(x)^2}\,dx = \int_{-1/2}^{\sqrt{3}/2} \dfrac{1}{\sqrt{1 - x^2}}\,dx = \arcsin(\sqrt{3}/2) - \arcsin(-1/2) = \dfrac{\pi}{2}$.

41. length $= \int_1^2 \sqrt{1 + (2x)^2}\,dx = \dfrac{1}{4}\left(4\sqrt{17} + \ln\left(4 + \sqrt{17}\right) - 2\sqrt{5} - \ln\left(2 + \sqrt{5}\right)\right)$
$$\approx 3.1678.$$

43. length $= \int_a^b \sqrt{1 + f'(x)^2}\,dx = \int_1^4 \sqrt{1 + \dfrac{9x}{4}}\,dx = \left(80\sqrt{10} - 13\sqrt{13}\right)/27 \approx 7.63371$.

45. Using the substitution $u = \sqrt{x}$ we get
$$\text{length} = \int_a^b \sqrt{1 + f'(x)^2}\,dx = \int_1^{16} \sqrt{1 + \sqrt{x}}\,dx = \dfrac{-8\sqrt{2}}{15} + \dfrac{40\sqrt{5}}{3} \approx 29.06.$$

47. length $= \int_a^b \sqrt{1 + f'(x)^2}\,dx = \int_3^9 \sqrt{\dfrac{1}{2} + \dfrac{1}{16x^2} + x^2}\,dx = 36 + \dfrac{\ln(3)}{4} \approx 36.2747$.

49. length $= \int_a^b \sqrt{1 + f'(x)^2}\,dx = \int_{1/2}^2 \sqrt{\dfrac{1}{2} + \dfrac{1}{16x^4} + x^4}\,dx = \int_{1/2}^2 \left(x^2 + \dfrac{1}{4x^2}\right) dx = 3$.

51. length $= \int_a^b \sqrt{1 + f'(x)^2}\, dx = \int_1^3 \sqrt{\frac{1}{2} + \frac{1}{x^6} + \frac{x^6}{16}}\, dx = \int_1^3 \left(\frac{1}{x^3} + \frac{x^3}{4} \right) dx = \frac{49}{9}.$

53. length $= \int_a^b \sqrt{1 + f'(x)^2}\, dx = \int_1^4 \sqrt{1 + \sinh^2 x}\, dx$

$$= \int_1^4 \cosh x\, dx = \sinh 4 - \sinh 1 \approx 26.1147.$$

55. The length of the curve $y = f(x)$ from $x = a$ to $x = b$ is $I = \int_a^b \ell(x)\, dx$, where
$\ell(x) = \sqrt{1 + \left(f'(x) \right)^2}$. Now $\ell'(x) = f'(x) f''(x) / \sqrt{1 + \left(f'(x) \right)^2}$. Because $f'(x) > 0$ and
$f''(x) < 0$, we see that $\ell'(x) < 0$ for $a \le x \le b$. Thus, $\ell(x)$ is decreasing on $[a, b]$; hence,
L_n *over*estimates the value of I.

57. After 5 minutes, the bug has crawled 30 feet. Since the length of the curve $y = \frac{1}{3} \left(x^2 + 2 \right)^{3/2}$
between $x = 1$ and $x = s$ is $s^3/3 + s - 4/3$, the x-coordinate of the bug's position after
5 minutes is $x \approx 4.3271$. Thus, the bug is (approximately) at the point $(4.3271, 31.4470)$.

59. The fundamental theorem of calculus (FTC) says that $f'(x) = \sqrt{g'(t)^2 - 1}$. Thus
$\sqrt{1 + f'(x)^2} = g'(t)$, and the desired arclength is (using the FTC again)
$I = \int_a^b g'(t)\, dt = g(b) - g(a)$.

§7.2 Finding Volumes by Integration

1. Nick = Rick $\cdot \pi/4 \approx 12.47$.

3. The cone has bottom circumference 4 meters, so bottom radius is $2/\pi$. Also, the entire cone has height 40 meters, so its volume is $\pi r^2 h/3 = \pi \cdot 4/\pi^2 \cdot 40/3 = 16.98$ cubic meters. The log part is the bottom half of the cone, which has volume 7/8 of entire cone, or about 14.85 cubic meters.

5. (a) The volume of a square cylinder with side s and height h is $s^2 h$; the Fact implies that a pyramid has 1/3 this volume.

   (b) Let $A(y)$ be the cross-sectional (square) area of the pyramid at height h. The side length at height y decreases linearly from s (when $y = 0$) to 0 (when $y = h$). This radius is therefore $s(1 - y/h)$, and so the cross-sectional area at height y is $s^2(1 - y/h)^2$. Integrating gives $\int_0^h s^2(1 - y/h)^2 \, dy = s^2 \int_0^h (1 - y/h)^2 \, dy = \dfrac{s^2 h}{3}$.

7. The cone's cross sectional area at height y is $\pi r^2(1 - y/h)^2$. Since the cone's total volume is $\pi r^2 h/3$, we want H so that $\pi r^2 \int_0^H (1 - y/h)^2 \, dy = \pi r^2 h/6$. Solving this equation for H gives $h(1 - 1/\sqrt[3]{2})$.

9. (a) The given sum is a Riemann sum for the volume integral $V = \int_a^b \pi f(x)^2 \, dx$.

   (b) If $f(x)$ is decreasing, then so is the integrand $\pi f(x)^2$ in the volume integral. Therefore, all right sums underestimate.

   (c) If $f(x) = x^2$, $a = 0$, and $b = 2$, then the given sums converge to the integral $V = \int_0^2 \pi x^4 \, dx$, which has value $32\pi/5$.

11. $V = \displaystyle\int_0^8 \pi \left(x^3\right)^2 dx = \dfrac{8^7 \pi}{7}$.

13. $V = \displaystyle\int_0^2 \pi \left((x + 6)^2 - (x^3)^2\right) dx = \dfrac{1688\pi}{21}$.

15. $V = \displaystyle\int_0^2 \pi \left(4^2 - (y^2)^2\right) dy = \dfrac{128\pi}{5}$.

17. $V = \displaystyle\int_0^4 \pi \left(\sqrt{y}\right)^2 dy - \int_1^4 \pi \left(\log_2 y\right)^2 dy = \pi \left(\dfrac{16}{\ln 2} - 8 - \dfrac{6}{(\ln 2)^2}\right) \approx 8.15214$.

19. For x in $[-2, 2]$, the cross-section at x is a square with edge-length $2\sqrt{4 - x^2}$; this square has area $4(4 - x^2)$. Thus the solid's volume is $4 \int_{-2}^2 (4 - x^2) \, dx = 128/3$.

21. Recall that an equilateral triangle of base s has area $\sqrt{3}s^2/4$. For x in $[0, 3]$, the cross-section at x is an equilateral triangle with base $s = 2\sqrt{3x}$. This triangle has area $3\sqrt{3}x$, so the solid's volume is $3\sqrt{3} \int_0^3 x \, dx = 27\sqrt{3}/2$.

23. (a) Circular cross sections all have area πr^2, so the volume is $\int_0^h \pi r^2 \, dx = \pi r^2 h$.

(b) Think of the cylinder as having its base inside the circle $x^2 + y^2 = r^2$ in the xy-plane. Cross sections perpendicular to the x-axis are then rectangles, with base $2\sqrt{r^2 - x^2}$ and height h, for $-r \le x \le r$. Thus the volume in question is $\int_{-r}^{r} 2\sqrt{r^2 - x^2}h \, dx = \pi r^2 h$.

25. The radius of the glass (in inches) at height h (in inches) is $r = 1 + 0.1h$ when $0 \le h \le 5$. Thus, the volume of the glass is $V = \pi \int_0^5 (1 + 0.1h)^2 \, dh = \dfrac{95\pi}{12} \approx 24.87$ in^3.

27. From the information given, the radius of the Earth is $r = C/2\pi \approx 3{,}963$ miles.

(a) $V = \pi \int_{\sqrt{2}r/2}^{r} (r^2 - y^2) \, dy = \left(\frac{2}{3} - \frac{5\sqrt{2}}{12} \right) \pi r^3 \approx 1.1514 \times 10^{10}$ miles^3.

(b) $V = \pi \int_0^{\sqrt{2}r/2} (r^2 - y^2) \, dy = \frac{5\sqrt{2}\pi r^3}{12} \approx 1.1522 \times 10^{11}$ miles^3.

29. **Method 1:** Cross-sections parallel to the ends of the tank have constant area:

$$A(x) = 2 \int_{-4}^{2} \sqrt{4^2 - y^2} \, dy = 2 \left(\frac{1}{2} y \sqrt{4^2 - y^2} + 8 \arcsin \left(\frac{y}{4} \right) \right) \Bigg]_{-4}^{2} = 4\sqrt{3} + \frac{32\pi}{3}.$$

Thus, the volume of gasoline in the tank is $V = \int_0^{25} A(x) \, dx = 100\sqrt{3} + \dfrac{800\pi}{3} \approx 1011.0$ cubic feet.

Method 2: The cross-section parallel to the surface at height y has area $A(y) = 25 \cdot 2\sqrt{4^2 - y^2}$. Thus, the volume of gasoline in the tank is

$$V = \int_{-4}^{2} A(y) \, dy = 50 \int_{-4}^{2} \sqrt{4^2 - y^2} \, dy.$$

Method 3: The volume of the tank is 400π cubic feet. The volume of the "empty" portion at the top is $25 \int_{-\sqrt{12}}^{\sqrt{12}} \left(\sqrt{4^2 - x^2} - 2 \right) dx$ cubic feet. Now the volume of the gasoline in the tank can be found by subtraction.

31. The volume is the integral $\int_0^{30} A(h) \, dh$, where $A(h)$ is the leg's cross-sectional area at height h; these areas can be found by squaring the width at each height. Thus we can find $A(0), A(5), \ldots A(30)$. Using these values gives the integral approximations $L_6 = 39.8$ $R_6 = 27.8$; $T_6 = 33.8$; $M_3 = 33.2$, all in cubic inches. (Alternatively, one might use the linear formula $w = 1.6 - 1.2h/30$ for width as a function of height; squaring this gives a quadratic formula for area, which can be integrated.)

33. Cross-sections parallel to the ends of the tank have area:

$$A(x) = 2 \int_{-6}^{3} \sqrt{9 - \tfrac{y^2}{4}} \, dy = 4 \int_{-3}^{3/2} \sqrt{9 - u^2} \, du$$

$$= 2 \left(u\sqrt{9 - u^2} + 9 \arcsin \left(\tfrac{u}{3} \right) \right) \Big|_{-3}^{3/2} = \frac{9\sqrt{3}}{2} + 12\pi.$$

Thus, the volume of fuel oil in the tank is $V = \int_{0}^{10} A(x) \, dx = 45\sqrt{3} + 120\pi \approx 454.93$ cubic feet.

35. (a) The rectangle has base $(b - a)$ and height h, so the area is $(b - a)h$.

    (b) Rotating R about the x-axis produces a circular cylinder of radius h and length $(b - a)$, so the volume is $\pi h^2 (b - a)$.

    (c) Rotating R about the line $y = -c$ produces a hollow tube with length $(b - a)$, outer radius $c + h$, and inner radius c. Thus the volume is the
$$\pi (c + h)^2 (b - a) - \pi c^2 (b - a) = \pi (2hc + h^2)(b - a).$$

    (d) Rotating R about the line $y = c$, where $c \geq h$, produces a hollow tube with length $(b - a)$, outer radius c, and inner radius $c - h$. Thus the volume is the
$$\pi c^2 (b - a) - \pi (c - h)^2 (b - a) = \pi (2hc - h^2)(b - a).$$

37. $V = \int_{0}^{1} \pi \left(1^2 - \left(1 - \sqrt{x} \right)^2 \right) dx = \frac{5\pi}{6}.$

39. $V = \int_{-4}^{0} \pi \left(\left(2 + \sqrt{x + 4} \right)^2 - \left(2 - \sqrt{x + 4} \right)^2 \right) dx$

$$+ \int_{0}^{5} \pi \left(\left(2 + \sqrt{x + 4} \right)^2 - (2 - (2 - x))^2 \right) dx$$

$$= \frac{128\pi}{3} + \frac{123\pi}{2} = \frac{625\pi}{6}.$$

41. Note that $y = \arctan x$ implies $x = \tan y$. Thus,

$$V = \pi \int_{0}^{\pi/4} \left(1 - \tan^2 y \right) dy = \pi \left(y - (\tan y - y) \right) \Big]_{0}^{\pi/4} = \pi \left(\frac{\pi}{2} - 1 \right) \approx 1.7932.$$

43. The solid is a "thick-walled cylinder," with outer radius b, inner radius a, and height h. Thus, its volume is $\pi b^2 h - \pi a^2 h$.

45. The volume formula $V = 2\pi \int_{a}^{b} x f(x) \, dx$ gives $2\pi \int_{0}^{\pi} x \sin x \, dx = 2\pi^2$. (Use the table of integrals to find an antiderivative.)

47. The shell method gives $V = 2\pi \int_{0}^{2} x(1 - x/2) \, dx = 4\pi/3$. (The problem can also be done in other ways.)

49. The shell method gives $V = 2\pi \int_1^e x \ln x \, dx = \pi \dfrac{e^2 + 1}{2}$.

51. The remaining solid can be formed by revolving the region bounded by $y = 0$, $x = a$, and $y = h - hx/r$ about the y-axis. The shell method gives
$$V = 2\pi \int_a^r x h(1 - x/r) \, dx = -\pi a^{2h} + \frac{2\pi a^3 h}{3r} + \frac{\pi h r^2}{3}.$$

53.   (a) Solving the equation $y = x\sqrt{1 - x^2}$ for x^2 yields $x^2 = \left(1 \pm \sqrt{1 - 4y^2}\right)/2$. Thus, the outer boundary of the region is the curve $x = \left(\dfrac{1 + \sqrt{1 - 4y^2}}{2}\right)^{1/2}$ and the inner boundary of the region is the curve $x = \left(\dfrac{1 - \sqrt{1 - 4y^2}}{2}\right)^{1/2}$ when $0 \le y \le 1/2$. Therefore, the volume of the solid of revolution is
$$V = \int_0^{1/2} \frac{\pi}{2}\left(1 + \sqrt{1 - 4y^2}\right) dy - \int_0^{1/2} \frac{\pi}{2}\left(1 - \sqrt{1 - 4y^2}\right) dy.$$

  (b) Using the method of cylindrical shells, $V = \int_0^1 \pi x f(x) \, dx = \int_0^1 2\pi x^2 \sqrt{1 - x^2} \, dx$.

  (c) The integral in part (b) can be evaluated using the substitutions $u = 1 - x^2$, $w = u - 1/2$ and the table of integrals:
$$\int_0^1 2\pi x^2 \sqrt{1 - x^2} \, dx = \pi \int_0^1 \sqrt{(1/2)^2 - (u - 1/2)^2} \, du$$
$$= \pi \int_{-1/2}^{1/2} \sqrt{(1/2)^2 - w^2} \, dw = \frac{\pi^2}{8}.$$

55.   (a) Let $p(r)$ be the population density at a distance r from the center of the city. Since p is a linear function it can be written in the form $p(r) = ar + b$. Now, $p(0) = K$ and $p(R) = 0$, so $b = K$ and $a = -K/R$. Thus, $p(r) = -Kr/R + K = K(1 - r/R)$.

  (b) The population of the city is $\int_0^R p(r) \, dr = \dfrac{KR}{2}$.

§7.3 Work

1. Let x denote the number of inches of compression.

   (a) To find k, use the equation (implicit in the problem statement) $F(2) = 2k = 10$. It follows that $k = 5$ (and hence that $F(x) = 5x$).

   (b) Compressing from 16 inches to 12 inches means compressing from $x = 2$ to $x = 6$. Thus the work done is

   $$W = \int_2^6 F(x)\, dx = \int_2^6 5x\, dx = 80 \text{ inch-pounds.}$$

3. (a) work $= \displaystyle\int_0^{10} kx\, dx = 50k$ ft-lbs.

   (b) work $= \displaystyle\int_a^{a+10} kx\, dx = k(50 + 10a)$ ft-lbs.

5. work $= \displaystyle\int_0^{200} 5 \cdot (200 - x)\, dx = 100{,}000$ ft-lbs.

7. For reasons as in Example 5, the necessary forces F_1 and F_2 for the two buckets are given, respectively, by

   $$\begin{aligned} F_1(x) &= 60 + 0.25 \cdot (60 - x) = 75 - \frac{x}{4}; \\ F_2(x) &= 50 + 0.25 \cdot (70 - x) = 67.5 - \frac{x}{4}. \end{aligned}$$

   (The unit of force is pounds; x measures distance in feet from the bottom of the well.) To find the work in each case, we integrate:

   $$\begin{aligned} W_1 &= \int_0^{60} F_1(x)\, dx = \int_0^{60} \left(75 - \frac{x}{4}\right) dx = 4050 \text{ ft-lbs;} \\ W_2 &= \int_0^{70} F_2(x)\, dx = \int_0^{70} \left(67.5 - \frac{x}{4}\right) dx = 4112.5 \text{ ft-lbs.} \end{aligned}$$

   Raising the *second* bucket takes a little more work.

9. Density ρ means that the fluid's mass is ρ times its volume. To solve the problem we imitate Example 6. Imagine the tank as the solid formed by revolving the lower half of the circle $x^2 + y^2 = r^2$ around the y-axis. At each height y with $-r \le y \le 0$, the thin slab at height y, with thickness Δy, has area $\pi x^2 = \pi(r^2 - y^2)$, volume $\pi(r^2 - y^2)\Delta y$, and mass $\rho\pi(r^2 - y^2)\Delta y$. The force needed to raise the slab is $g\rho\pi(r^2 - y^2)\Delta y$, where g is the gravitational constant. The slab moves through distance $-y$, so the work done on the slab is $-y \cdot g\rho\pi(r^2 - y^2)\Delta y$. Integrating over y gives the answer:

   Work $= g\rho\pi \displaystyle\int_{-r}^0 -y(r^2 - y^2)\, dy = \dfrac{\pi\rho g r^4}{4}$. (The units of the answer depend on the units given for the data.)

11.   work $= \displaystyle\int_{-4}^{4} 42 \cdot 2\sqrt{16 - y^2} \cdot 15 \cdot (y + 17)\, dy$

$= -420\left(16 - y^2\right)^{3/2} + 10{,}710y\sqrt{16 - y^2} + 171{,}360\arcsin(y/4)\Big|_{-4}^{4}$

$= 171{,}360\pi$ ft-lbs $\approx 538{,}343$ ft-lbs.

13.   Let x denote the distance (in feet) that the spring is extended. Notice that since the chain weighs 20 pounds, we start with $x = 5$. (Draw a picture! Note, too, that other x-scales are possible.)

The problem is to find $\int_{5}^{7} F(x)\, dx$, where $F(x)$ is the net downward force necessary at a given x.

Let's find a formula for $F(x)$; there are two main ingredients: (i) For any value of x, the spring exerts an *upward* force of $4x$ pounds. (ii) For a given value of x, the length of chain remaining above the floor is $10 - (x - 5) = 15 - x$ feet. (A diagram should make this convincing.) Since the chain weighs 2 pounds per foot, this length of chain exerts a *downward* force of $2(15 - x) = 30 - 2x$ pounds.

Putting (i) and (ii) together means that the *net* downward force required for given x is $F(x) = 4x - (30 - 2x) = 6x - 30$ pounds. Thus the desired work is

$$W = \int_{5}^{7} F(x)\, dx = \int_{5}^{7} (6x - 30)\, dx = 12 \text{ ft-lbs.}$$

15.   The work is the signed area under the graph of F from $x = a$ to $x = b$.

17.   (a)   The expression $\displaystyle\int_{0}^{h} A(y)\, dy$ represents the volume of the tank.

(b)   The slab at height y gets lifted a distance of $h - y$, so the needed work (using foot-pounds as units) is $62.4 \int_{0}^{a} A(y)(h - y)\, dy$.

§7.4 Separating Variables: Solving DEs Symbolically

1. (a) $200\,°F$; $160\,°F$; $120\,°F$; $70\,°F$; $40\,°F$.

    (b) The value of T_0 is the temperature of the coffee at time $t = 0$. Thus, the figure shows the cooling curves for cups of coffee that start at each of the temperatures listed in part (a).

3. $y' = -0.1(y - 70)$, $y(0) = 70$.

5. (a) One way to find the values of T_r that correspond to C_1, C_4, and C_9 is just to observe that T_r represents the "long-run" temperature—the temperature to which the coffee tends over a long period of time. Looking at the right-hand parts of the graphs lets us read off these numbers for C_1 and C_4. Thus C_1 corresponds to $T_r = 100$; C_4 corresponds to $T_r = 70$. Not enough of C_9 appears for this to work very well. Instead, we can use the fact that for C_9, $y(10) = 70$. From this it follows:

    $$y(10) = (190 - T_r)e^{-1} + T_r = 70 \implies T_r \approx 0.$$

    (b) The value of T_r is the temperature of the room. Thus, the values found in part (a) give the temperature of the room in which each cup of coffee cooled.

7. For C_9, $y(10) = 70$. Therefore, $y(10) = (190 - T_r)e^{-1} + T_r = 70 \implies T_r \approx 0$. Thus, the IVP corresponding to curve C_9 is $y' = -0.1y$; $y(0) = 190$.

9. We need to check (1) that $y(t) = (T_0 - T_r)e^{-0.1t} + T_r$ solves the DE $y' = -0.1(y - T_r)$; and (2) that $y(0) = T_0$. The latter is easy: $y(0) = (T_0 - T_r)e^0 + T_r = T_0$, as claimed.

    To check (1), we calculate both sides of the DE and see that they agree:

    $$\begin{aligned} y(t)' &= (T_0 - T_r) \cdot e^{-0.1t} \cdot (-0.1); \\ -0.1(y - T_r) &= (-0.1) \cdot (T_0 - T_r)e^{-0.1t}. \end{aligned}$$

11. If $T_r = 80$ and $y(10) = 120$, then
    $$y(10) = (T_0 - 80)e^{-1} + 80 = 120 \implies T_0 = -\left(-\tfrac{80}{e} - 40\right)e \approx 188.731.$$

13. (a) Since $\lim_{t \to \infty} e^{-0.1t} = 0$, $\lim_{t \to \infty} y(t) = \lim_{t \to \infty}(T_0 - T_r)e^{-0.1t} + T_r = T_r$.

    (b) In coffee language, this means (as experience suggests!) that in the long run coffee cools to room temperature.

15. Differentiating the right side of $P' = kP(C - P)$ with respect to P gives

    $$\frac{d}{dP}(kP(C - P)) = \frac{d}{dP}\left(kCP - kP^2\right) = kC - 2kP = 0 \iff P = \frac{C}{2}.$$

    Thus P' has its maximum value where $P = C/2$, as claimed. (In rumor language: The rumor spreads fastest when *half* the people know it.)

17. $y' = y^2 \implies y'/y^2 = 1 \implies -1/y = x + C$ so $y = -1/(x + C)$. The initial condition $y(1) = 1$ implies that $C = -2$. Thus, the solution of the IVP is $y = -1/(x - 2)$.

19. $y' = x^2 y \implies y'/y = x^2 \implies \ln y = x^3/3 + C$ so $y = Ae^{x^3/3}$. Thus, the solution of the IVP is $y = -2e^{x^3/3}$.

21. $y' - xy^2 = 0 \implies y'/y^2 = x \implies -1/y = x^2/2 + C$ so $y = 2/(A - x^2)$. Thus, the solution of the IVP is $y = 2/(3 - x^2)$.

23.   (a) $f(y) = 1/y$, $g(x) = x$.

   (b) $F(y) = \ln y$, $G(x) = x^2/2$.

   (c) Differentiating both sides of the equation $F(y) = G(x) + C$ with respect to x leads to the equation $F'(y)y' = G'(x)$. Since $F' = f$ and $G' = g$, this equation is equivalent to the equation $f(y)y' = g(x)$. Thus, if y is implicitly defined by the equation $F(y) = G(x) + C$, y is a solution of the DE $y' = xy$.

   (d) Let $F(y) = G(x) + C$, where C is a constant. Then, using the results from parts (a)–(c), we have

$$\ln y = \frac{x^2}{2} + C \implies y = e^{x^2/2+C} = Ke^{x^2/2}.$$

25.   (a) The exact values of k for curves P_1 to P_4 are, respectively, 2, 1.5, 1, and 0.5. (Thus the values of $-k$ are, respectively, -2, -1.5, -1, and -0.5.)

   The k-values can be found—approximately—by reading the graphs. The graph of P_1, for example, seems to pass through the point $(2, 750)$. This gives an equation we can solve for k, as follows:

$$P_1(2) = \frac{1000}{19e^{-2k} + 1} \approx 750 \implies k = \frac{\ln 57}{2} \approx 2.02.$$

   Values of k for the other curves are found similarly.

   (b) P_1 corresponds to the "hottest" rumor, P_4 to the "coolest." The graph shapes agree—hotter rumors spread faster.

27. $\arctan y = x + C$ or $y = \tan(x + C)$.

29. $\dfrac{1}{4} \ln \left| \dfrac{y + 2}{y - 2} \right| = x + C$ or $y = \dfrac{2Ae^{4x} + 2}{Ae^{4x} - 1}$.

31. $\ln(\sec y + \tan y) = x^3/3 + C$.

§7.5   Present Value

1. For any interest rate r, the present value of one \$1 million payment, 23 years ahead, is
   $PV = 1,000,000 \cdot e^{-r \cdot 23}$.

   (a) If $r = 0.06$; then $PV = \$1,000,000e^{-0.06 \cdot 23} \approx \$251,579$.

   (b) If $r = 0.08$, then $PV = \$1,000,000e^{-0.08 \cdot 23} \approx \$158,817$.

   (c) To find the desired r we solve the equation $PV = 100,000 = 1,000,000e^{-r \cdot 23}$ for r.
       The result: $r = (\ln 10)/23 \approx 0.10011$—just a bit above 10%.

3. For any interest rate r (real or nominal) the present value of one \$1 million payment, 23
   years ahead, is $PV = 1,000,000 \cdot e^{-r \cdot 23}$.

   (a) If $r = 0.02$, then $PV = \$1,000,000e^{-0.02 \cdot 23} \approx \$631,284$.

   (b) If $r = 0.04$, then $PV = \$1,000,000e^{-0.04 \cdot 23} \approx \$398,519$.

   (c) To find Betty's r we solve the equation $PV = 200,000 = 1,000,000e^{-r \cdot 23}$ for r. The
       result: $r = (\ln 5)/23 \approx 0.07$. Thus Betty needs to find a *real* interest rate—after
       inflation—of 7%.

5. At any interest rate r, the present value formula for several future payments says, in this
   situation, that

   $$PV = 40,000e^{-18r} + 42,000e^{-19r} + 44,000e^{-20r} + 46,000e^{-21r}.$$

   If $r = 0.06$, the formula (and some electronic help) give $PV \approx \$53,316.85$. If $r = 0.08$,
   $PV \approx \$36,119.66$.

7. (a) p_I is the graph which peaks at $t = 180$. When $t = 180$, $p_I = 110$.

   (b) The function $\cos t$ has period 2π. The functions p_T and p_I both have period 360.

   (c) The constant 50 affects the amplitude of the graphs. The constant 60 shifts the graphs
       upward. The constants 180, 105, and π affect where each graph has its local maxima
       and minima.

9. The total return from the investment is \$40,000 paid continuously over an 8-year time
   interval. Since the present value of the return from the investment is
   $PV = \$5,000 \int_0^8 e^{-0.06t} \, dt \approx \$31,768$, this is a worthwhile investment.

11. (a) $\left(\dfrac{e^{ax}}{a^2 + b^2} \left(a \cos(bx) + b \sin(bx) \right) \right)' = \dfrac{a e^{ax}}{a^2 + b^2} \left(a \cos(bx) + b \sin(bx) \right)$

   $$+ \frac{e^{ax}}{a^2 + b^2} \left(-ab \sin(bx) + b^2 \cos(bx) \right)$$

   $$= e^{ax} \cos(bx).$$

(b) $\displaystyle\int_0^{360} \left(50\cos\left(\pi \cdot \frac{t-180}{180} \right) + 60 \right) e^{-0.1t/360}\, dt$

$$= \left(216{,}000 - \frac{180{,}000}{1+400\pi^2} \right) \left(1 - e^{-1/10} \right)$$

$$\approx 20{,}550.78.$$

NOTE: $\displaystyle\cos\left(\pi \cdot \frac{t-180}{180} \right) = \cos\left(\frac{\pi t}{180} - \pi \right) = -\cos\left(\frac{\pi t}{180} \right).$

13. The interest earned from the income received at time t is $p(t)e^{r(T-t)}$. Thus, the total income accrued between $t = 0$ and $t = T$ is $\displaystyle\int_0^T p(t)e^{r(T-t)}\, dt = e^{rT} \int_0^T p(t)e^{-rt}\, dt = e^{rT}\, PV.$

8   Symbolic Antidifferentiation Techniques

§8.1   Integration by Parts

1. (a) Using the product rule, $(x \ln x)' = x \cdot 1/x + 1 \cdot \ln x = 1 + \ln x.$

    (b) Part (a) implies that $\int (1 + \ln x)\, dx = x + \int \ln x\, dx = x \ln x + C.$ Thus,

    $$\int \ln x\, dx = x \ln x - x + C.$$

3. (a) Using the product and chain rules, $(xe^{-x})' = 1 \cdot e^{-x} + x \cdot (-1)e^{-x} = e^{-x} - xe^{-x}.$

    (b) Part (a) implies that $\int (e^{-x} - xe^{-x})\, dx = -e^{-x} - \int xe^{-x}\, dx = xe^{-x} + K.$ Thus,

    $$\int xe^{-x}\, dx = -e^{-x} - xe^{-x} + C.$$

    (c) If $u = x$ and $dv = e^{-x}\, dx$, then $du = dx$ and $v = -e^{-x}$. Thus, using integration by parts,

    $$\int xe^{-x}\, dx = x \cdot (-e^{-x}) - \int (-e^{-x})\, dx = -xe^{-x} + \int e^{-x}\, dx = -xe^{-x} - e^{-x} + C.$$

5. (a) Using the product and chain rules,

    $$\left(x \arctan(2x)\right)' = 1 \cdot \arctan(2x) + x \cdot \frac{2}{1 + (2x)^2} = \arctan(2x) + \frac{2x}{1 + 4x^2}.$$

    (b) Part (a) implies that

    $$\int \left(\arctan(2x) + \frac{2x}{1 + 4x^2} \right) dx = \int \arctan(2x)\, dx + \frac{1}{4} \ln(1 + 4x^2)$$
    $$= x \arctan(2x) + C.$$

    Thus, $\int \arctan(2x)\, dx = x \arctan(2x) - \dfrac{1}{4} \ln(1 + 4x^2) + C.$

    (c) If $u = \arctan(2x)$ and $dv = dx$, then $du = (2\, dx)/(1 + 4x^2)$ and $v = x$. Thus, using integration by parts,

    $$\int \arctan(2x)\, dx = x \arctan(2x) - \int \frac{2x\, dx}{1 + 4x^2} = x \arctan(2x) - \frac{1}{4} \ln(1 + 4x^2) + C.$$

7. (a) Using the product and chain rules,
    $$(e^{2x} \sin x)' = 2e^{2x} \cdot \sin x + e^{2x} \cdot \cos x = (2 \sin x + \cos x)e^{2x}.$$

(b) Using the product and chain rules,

$$(e^{2x}\cos x)' = 2e^{2x}\cdot\cos x + e^{2x}\cdot(-\sin x) = (2\cos x - \sin x)e^{2x}.$$

(c) Part (a) implies that

$$\int (2\sin x + \cos x)e^{2x}\,dx = 2\int e^{2x}\sin x\,dx + \int e^{2x}\cos x\,dx = e^{2x}\sin x + K,\text{ so}$$

$$2\int e^{2x}\sin x\,dx = e^{2x}\sin x - \int e^{2x}\cos x\,dx + K.\text{ Now, part (b) implies that}$$

$$\int e^{2x}\sin x\,dx = 2\int e^{2x}\cos x\,dx - e^{2x}\cos x.\text{ Combining these results,}$$

$$2\cdot 2\int e^{2x}\sin x\,dx + \int e^{2x}\sin x\,dx = 2e^{2x}\sin x - 2\int e^{2x}\cos x\,dx + 2K$$

$$+ 2\int e^{2x}\cos x\,dx - e^{2x}\cos x$$

$$= e^{2x}(2\sin x - \cos x) + 2K.$$

Thus, $\displaystyle\int e^{2x}\sin x\,dx = \frac{e^{2x}}{5}(2\sin x - \cos x) + C.$

(d) Let $u = \sin x$ and $dv = e^{2x}\,dx$. Then, $du = \cos x\,dx$ and $v = e^{2x}/2$ so

$$\int e^{2x}\sin x\,dx = \frac{e^{2x}}{2}\sin x - \frac{1}{2}\int e^{2x}\cos x\,dx.$$

Now, if $u = \cos x$ and $dv = e^{2x}\,dx$, $du = -\sin x\,dx$ and $v = e^{2x}/2$ so

$$\int e^{2x}\cos x\,dx = \frac{e^{2x}}{2}\cos x + \frac{1}{2}\int e^{2x}\sin x\,dx.$$

Thus, $\displaystyle\int e^{2x}\sin x\,dx = \frac{e^{2x}}{2}\sin x - \frac{1}{2}\left(\frac{e^{2x}}{2}\cos x + \frac{1}{2}\int e^{2x}\sin x\,dx\right)$

$$\implies \frac{5}{4}\int e^{2x}\sin x\,dx = \frac{e^{2x}}{4}(2\sin x - \cos x) + C$$

$$\implies \int e^{2x}\sin x\,dx = \frac{e^{2x}}{5}(2\sin x - \cos x) + C.$$

9. $\displaystyle\int x\sin(3x)\,dx = \frac{\sin(3x)}{9} - \frac{x\cos(3x)}{3} + C \quad [du = dx,\, v = -\tfrac{1}{3}\cos(3x)].$

11. $\displaystyle\int x\sec^2 x\,dx = x\tan x + \ln|\cos x| + C \quad [du = dx,\, v = \tan x].$

13. $\displaystyle\int (\ln x)^2\,dx = (\ln x)(x\ln x - x) - \int (x\ln x - x)\frac{dx}{x}$

$$= (\ln x)(x\ln x - x) - (x\ln x - x) + x + C$$

$$= x(\ln x)^2 - 2x\ln x + 2x + C \quad [du = dx/x,\, v = x\ln x - x].$$

15. $\displaystyle\int_0^1 xe^{-x}\,dx = -(1+x)e^{-x}\Big]_0^1 = 1 - 2e^{-1} \approx 0.26424.$ $[M_{20} \approx 0.26435]$.

17. $\displaystyle\int_1^e x\ln x\,dx = \frac{1}{2}x^2\ln x - \frac{1}{4}x^2\Big]_1^e = \frac{1}{4}\left(e^2+1\right) \approx 2.09726.$ $[M_{20} \approx 2.0970]$.

19. $\displaystyle\int_{-1}^{\sqrt{2}/2} x^2\arctan x\,dx = \frac{1}{3}x^3\arctan x - \frac{1}{6}x^2 + \frac{1}{6}\ln\left(1+x^2\right)\Big]_{-1}^{\sqrt{2}/2}$

$\displaystyle = \frac{\sqrt{2}}{12}\arctan(\sqrt{2}/2) + \frac{1}{12} + \frac{1}{6}\ln(3/4) - \frac{\pi}{12} \approx -0.15388.$ $[M_{20} \approx -0.15361]$.

21. (a) If $u = x^2$ and $dv = \cos x\,dx$, then $du = 2x\,dx$ and $v = \sin x$. Thus, using integration by parts, $\displaystyle\int x^2\cos x\,dx = x^2\sin x - 2\int x\sin x\,dx$.

   (b) Let $u = x$ and $dv = \sin x\,dx$. Then, $du = dx$ and $v = -\cos x$ so

   $$\int x\sin x\,dx = -x\cos x + \int \cos x\,dx = -x\cos x + \sin x + C.$$

   (c) Combining parts (a) and (b),

   $$\int x^2\cos x\,dx = x^2\sin x + 2x\cos x - 2\sin x + C.$$

23. Since $n = 3$, repeated use of the reduction formula yields

$$\int (\ln x)^3\,dx = x\,(\ln x)^3 - 3\int (\ln x)^2\,dx$$

$$= x\,(\ln x)^3 - 3x\,(\ln x)^2 + 6\int \ln x\,dx$$

$$= x\,(\ln x)^3 - 3x\,(\ln x)^2 + 6x(\ln x - 1) + C.$$

25. Using integration by parts with $u = \ln x$, $dv = x^r\,dx$, $du = dx/x$ and $v = x^{r+1}/(r+1)$:

$$\int x^r\ln x\,dx = \frac{x^{r+1}\ln x}{r+1} - \frac{1}{r+1}\int x^{r+1}\frac{dx}{x} = \frac{x^{r+1}\ln x}{r+1} - \frac{1}{r+1}\int x^r\,dx$$

$$= \frac{x^{r+1}}{r+1}\left(\ln x - \frac{1}{r+1}\right) + C.$$

27. Let $u = (\ln x)^n$ and $dv = dx$. Then $du = n\,(\ln x)^{n-1}\,dx$ and $v = x$. Therefore,

$$\int (\ln x)^n\,dx = x\,(\ln x)^n - n\int (\ln x)^{n-1}\,dx, \quad n \geq 1.$$

29.   (a) If $u = x^3$, then $du = 3x^2\,dx$, so
$$\int x^2 \ln(x^3)\,dx = \frac{1}{3}\int \ln u\,du = \frac{u}{3}(\ln u - 1) = \frac{x^3}{3}\left(\ln(x^3) - 1\right).$$

   (b) Let $u = \ln x$, $dv = x^2\,dx$, $du = dx/x$, and $v = x^3/3$. Then, using integration by parts,
$$\int x^2 \ln(x^3)\,dx = 3\int x^2 \ln x\,dx = x^3 \ln x - \int x^2\,dx = x^3 \ln x - \frac{x^3}{3}.$$

31. $\displaystyle\int x^3 e^{x^2}\,dx = \int x \cdot x^2 e^{x^2}\,dx = \frac{1}{2}\int u e^u\,du$
$$= \frac{1}{2}\left(u e^u - \int e^u\,du\right) = (u - 1)e^u/2 = (x^2 - 1)e^{x^2}/2.$$

33.   (a) Let $u = \sin x$ and $dv = \sin x\,dx$. Then, $du = \cos x\,dx$ and $v = -\cos x$ so
$$\int \sin^2 x\,dx = -\sin x \cos x + \int \cos^2 x\,dx.$$

   (b) $\displaystyle\int \sin^2 x\,dx = -\sin x \cos x + \int \cos^2 x\,dx = -\sin x \cos x + \int \left(1 - \sin^2 x\right)\,dx.$

   Therefore, $2\displaystyle\int \sin^2 x\,dx = x - \sin x \cos x$ and $\displaystyle\int \sin^2 x\,dx = \frac{1}{2}(x - \sin x \cos x) + C.$

35.   Let $u = x$ and $dv = \cos^2 x\,dx = \frac{1}{2}\left(1 + \cos(2x)\right)\,dx$. Then $du = dx$ and
$v = \frac{1}{2}\left(x + \frac{1}{2}\sin(2x)\right)$. Therefore,
$$\int x \cos^2 x\,dx = \frac{1}{2}x^2 + \frac{1}{4}x \sin(2x) - \frac{1}{2}\int \left(x + \frac{\sin(2x)}{2}\right)\,dx$$
$$= \frac{1}{4}x^2 + \frac{1}{4}x \sin(2x) + \frac{1}{8}\cos(2x) + C.$$

37. $\displaystyle\int x^2 \cos x\,dx = x^2 \sin x - 2\sin x + 2x \cos x + C.$

   Let $u = x^2$, $dv = \cos x\,dx$. Then, $\displaystyle\int x^2 \cos x\,dx = x^2 \sin x - 2\int x \sin x\,dx$. The remaining
   antiderivative can be found using integration by parts again.

39.   Let $u = x$ and $dv = \csc x \cot x\,dx$. Then $du = dx$ and $v = -\csc x$ so
$$\int x \csc x \cot x\,dx = -x \csc x + \int \csc x\,dx = -x \csc x + \ln|\csc x - \cot x| + C.$$

41. $\displaystyle\int \sqrt{x}\,\ln\left(\sqrt[3]{x}\right)\,dx = \frac{1}{3}\int \sqrt{x}\,\ln x\,dx = \frac{2}{9}x^{3/2}\ln x - \frac{4}{27}x^{3/2} + C.$
   $[u = \ln x, dv = \sqrt{x}\,dx]$.

43. $\displaystyle\int \arctan(1/x)\,dx = x \arctan(1/x) + \frac{1}{2}\ln\left(x^2 + 1\right) + C.$ $[u = \arctan(1/x), dv = dx]$.

45. $\displaystyle\int e^{\sqrt{x}}\,dx = 2e^{\sqrt{x}}\sqrt{x} - 2e^{\sqrt{x}} + C.$

First substitute $w = \sqrt{x}, w^2 = x, 2w\,dw = dx$. This produces the new integral $\int 2we^{w}\,dw$. Now use parts, with $u = w, dv = e^{w}\,dw$. The answer follows directly.

47. $\displaystyle\int \sin(\ln x)\,dx = \frac{x}{2}\big(\sin(\ln x) - \cos(\ln x)\big) + C.$

First substitute $w = \ln x, x = e^{w}, dx = e^{w}\,dw$. This gives the new integral
$\int \sin(\ln x)\,dx = \int e^{w}\sin w\,dw$. The new integral is done (with integration by parts, twice) in the book.

49. $\displaystyle\int \sin\left(\sqrt{x}\right)\,dx = 2\sin\left(\sqrt{x}\right) - 2\sqrt{x}\cos\left(\sqrt{x}\right) + C.$

First substitute $w = \sqrt{x}$, then use integration by parts with $u = w$ and $dv = \sin w\,dw$.

51. $\displaystyle\int \sqrt{x}\arctan\left(\sqrt{x}\right)\,dx = \frac{2}{3}x^{3/2}\arctan\left(\sqrt{x}\right) - \frac{1}{3}(1+x) + \frac{1}{3}\ln(1+x) + C.$

Substitute $w = \sqrt{x}$, then use integration by parts with $u = \arctan w$ and $dv = w^2\,dw$.

53. First, make the substitution $u = \sin x$. Then,

$$\int \cos x\,\ln(\sin x)\,dx = \int \ln u\,du = u\ln u - u = (\sin x)\ln(\sin x) - \sin x + C.$$

55. Using integration by parts with $u = x$ and $dv = \sinh x\,dx$,

$$\int x\sinh x\,dx = x\cosh x - \int \cosh x\,dx = x\cosh x - \sinh x + C.$$

57. Using integration by parts (first with $u = x^2$ and $dv = \sinh x\,dx$, then with $u = x$ and $dv = \cosh x$),

$$\int x^2\sinh x\,dx = x^2\cosh x - 2\int x\cosh x\,dx = x^2\cosh x - 2x\sinh x + 2\cosh x + C.$$

59. (a) $\displaystyle I_1 = \int_1^e \ln x\,dx = \big(x\ln x - x\big)\Big|_1^e = (e - e) - (0 - 1) = 1.$

(b) Using the reduction formula,
$$I_2 = \int_1^e (\ln x)^2\,dx = x(\ln x)^2\Big|_1^e - 2\int_1^e \ln x\,dx = (e - 0) - 2I_1 = e - 2.$$

(c) Using the reduction formula,
$$I_3 = \int_1^e (\ln x)^3\,dx = x(\ln x)^3\Big|_1^e - 3\int_1^e (\ln x)^2\,dx = (e - 0) - 3I_2 = 6 - 2e.$$

(d) The reduction formula implies that $I_n = e - nI_{n-1}$. Thus, $I_4 = e - 4I_3 = 9e - 24$ and $I_5 = e - 5I_4 = 120 - 44e.$

61. $f(x) = x^4/4.$ Integration by parts with $u = \cos x$ and $dv = x^3\,dx$ leads to $\displaystyle\int x^3\cos x\,dx = $
$\frac{1}{4}x^4\cos x + \int \frac{1}{4}x^4\sin x\,dx.$

63. Using integration by parts (twice), we can show that

$$\int f(x)\cos x \, dx = f(x)\sin x - \int f'(x)\sin x \, dx$$

$$= f(x)\sin x + f'(x)\cos x - \int f''(x)\cos x \, dx.$$

Thus,

$$\int_{-\pi/2}^{3\pi/2} f(x)\cos x \, dx = \left(f(x)\sin x + f'(x)\cos x\right)\Big]_{-\pi/2}^{3\pi/2} - \int_{-\pi/2}^{3\pi/2} f''(x)\cos x \, dx$$

$$= -f(3\pi/2) + f(-\pi/2) - 4 = -\int_{-\pi/2}^{3\pi/2} f'(x)\, dx - 4 = -5.$$

65. (a) This is the fundamental theorem of calculus.

(b) Using integration by parts with $u = f'(x), dv = dx$, and $v = x - b$, leads to

$$\int_{a}^{b} f'(x)\, dx = (x - b)f'(x)\Big|_{a}^{b} - \int_{a}^{b} (x - b)f''(x)\, dx$$

$$= (b - a)f'(a) - \int_{a}^{b} (x - b)f''(x)\, dx.$$

Combining this result with part (a) implies that

$$f(b) = f(a) + f'(a)(b - a) - \int_{a}^{b} (x - b)f''(x)\, dx.$$

(c) Using integration by parts with $u = f''(x), dv = (x - b)dx$, and $v = (x - b)^2/2$, leads to

$$\int_{a}^{b} (x - b)f''(x)\, dx = \frac{1}{2}(x - b)^2 f''(x)\Big|_{a}^{b} - \frac{1}{2}\int_{a}^{b} (x - b)^2 f'''(x)\, dx$$

$$= -\frac{1}{2}(b - a)^2 f''(a) - \frac{1}{2}\int_{a}^{b} (x - b)^2 f'''(x)\, dx.$$

Combining this with part (b) implies that

$$f(b) = f(a) + f'(a)(b - a) + \frac{1}{2}f''(a)(b - a)^2 + \frac{1}{2}\int_{a}^{b} (x - b)^2 f'''(x)\, dx.$$

(d) Using integration by parts with $u = f'''(x), dv = (x - b)dx$, and $v = (x - b)^3/3$, leads to

$$\int_{a}^{b} (x - b)^2 f'''(x)\, dx = \frac{1}{3}(x - b)^3 f'''(x)\Big|_{a}^{b} - \frac{1}{3}\int_{a}^{b} (x - b)^3 f^{(4)}(x)\, dx$$

$$= \frac{1}{3}f'''(a)(b - a)^3 - \frac{1}{3}\int_{a}^{b} (x - b)^3 f^{(4)}(x)\, dx.$$

Thus,

$$f(b) = f(a) + f'(a)(b-a) + \frac{1}{2}f''(a)(b-a)^2 + \frac{1}{6}f'''(a)(b-a)^3$$

$$-\frac{1}{6}\int_a^b (x-b)^3 f^{(4)}(x)\, dx$$

$$= p_3(b) - \frac{1}{6}\int_a^b (x-b)^3 f^{(4)}(x)\, dx$$

so $f(b) - p_3(b) = -\dfrac{1}{6}\displaystyle\int_a^b (x-b)^3 f^{(4)}(x)\, dx.$

§8.2   Partial Fractions

1.   (a) $\dfrac{1}{1+x} + \dfrac{1}{1-x} = \dfrac{(1-x)+(1+x)}{(1+x)(1-x)} = \dfrac{2}{1-x^2}.$

   (b) $\displaystyle\int \dfrac{dx}{1-x^2} = \dfrac{1}{2}\int \left(\dfrac{1}{1+x} + \dfrac{1}{1-x}\right) dx$

   $= \dfrac{\ln|1+x|}{2} - \dfrac{\ln|1-x|}{2} + C = \dfrac{1}{2}\ln\left|\dfrac{1+x}{1-x}\right| + C.$

3. $\dfrac{x^3}{1+x^2} = \dfrac{x(1+x^2)-x}{1+x^2} = x - \dfrac{x}{1+x^2}.$

5.   (a) $\dfrac{2}{x+1} + \dfrac{3}{x+2} = \dfrac{2(x+2)+3(x+1)}{(x+1)(x+2)} = \dfrac{5x+7}{(x+1)(x+2)}.$

   (b) $\displaystyle\int \dfrac{5x+7}{(x+1)(x+2)}\, dx = 2\ln|x+1| + 3\ln|x+2| + C.$

7.   (a) f is the quotient of two polynomials.

   (b) Both the numerator and the denominator of the expression defining f have degree 2.

   (c) $f(x) = \dfrac{x^2+1}{(x-1)^2} = \dfrac{x^2+1-2x+2x}{x^2-2x+1} = 1 + \dfrac{2x}{x^2-2x+1}.$

   (d) $\displaystyle\int f(x)\,dx = \int\left(1 + \dfrac{2x}{x^2-2x+1}\right)dx = \int\left(1 + \dfrac{2x-2}{x^2-2x+1} + \dfrac{2}{(x-1)^2}\right)dx$

   $= x + \ln(x^2-2x+1) - \dfrac{2}{x-1} = x + 2\ln|x-1| - \dfrac{2}{x-1}.$

9.   Yes, because $x^2 + 2x + 3$ is an irreducible quadratic polynomial.

11.   No — $x^2 - 2x + 1 = (x-1)^2$ so the partial fraction decomposition of g has the form
   $\dfrac{A}{x+1} + \dfrac{B}{x-1} + \dfrac{C}{(x-1)^2}.$

13.   No — since $(x+1)^2$ is a repeated linear factor and $x^2 + 1$ is an irreducible quadratic factor, the partial fraction decomposition of h has the form $\dfrac{A}{x+1} + \dfrac{B}{(x+1)^2} + \dfrac{Cx+D}{x^2+1}.$

15.   (a) After multiplying, the result is $\dfrac{x^2+3x-1}{(x+1)(x-2)} = A + \dfrac{Bx}{x+1} + \dfrac{Cx}{x-2}.$ Substituting $x = 0$ into this equation yields $A = 1/2.$

   (b) After multiplying, the result is $\dfrac{x^2+3x-1}{x(x-2)} = \dfrac{A(x+1)}{x} + B + \dfrac{C(x+1)}{x-2}.$
   Substituting $x = -1$ into this equation yields $B = -1.$

   (c) After multiplying, the result is $\dfrac{x^2+3x-1}{x(x+1)} = \dfrac{A(x-2)}{x} + \dfrac{B(x-2)}{x+1} + C.$
   Substituting $x = 2$ into this equation yields $C = 3/2.$

(d) $\displaystyle\int \frac{x^2 + 3x - 1}{x(x+1)(x-2)}\, dx = \frac{1}{2}\int \left(\frac{1}{x} - \frac{2}{x+1} + \frac{3}{x-2}\right) dx$

$\displaystyle\qquad\qquad\qquad\qquad\quad = \bigl(\ln|x| - 2\ln|1+x| + 3\ln|x-2|\bigr)/2 + C.$

17. Since $\displaystyle\frac{x^2 + 2x + 5}{(x-1)(x+1)(x+2)} = \frac{4}{3(x-1)} - \frac{2}{x+1} + \frac{5}{3(x+2)}$,

$\displaystyle\int \frac{x^2 + 2x + 5}{(x-1)(x+1)(x+2)}\, dx = \frac{4\ln|x-1|}{3} - 2\ln|x+1| + \frac{5\ln|x+2|}{3} + C.$

19. (a) After multiplying, the result is $\displaystyle\frac{x^2-1}{x^2+4} = A + \frac{x(Bx+C)}{x^2+4}$. Substituting $x = 0$ into this equation yields $A = -1/4$.

(b) Using the result found in part (a), the partial fraction decomposition equation becomes $\displaystyle\frac{x^2-1}{x(x^2+4)} = -\frac{1}{4x} + \frac{Bx+C}{x^2+4}$. Substituting $x = 1$ into this equation yields $0 = -\dfrac{1}{4} + \dfrac{B+C}{5}$. Thus, $B + C = 5/4$.

(c) Using the result found in part (a), the partial fraction decomposition equation becomes $\displaystyle\frac{x^2-1}{x(x^2+4)} = -\frac{1}{4x} + \frac{Bx+C}{x^2+4}$. Substituting $x = -1$ into this equation yields $0 = \dfrac{1}{4} + \dfrac{C-B}{5}$. Thus, $B - C = 5/4$.

(d) $B = 5/4$ and $C = 0$ is the unique solution of the system equations $B + C = 5/4$ and $B - C = 5/4$.

(e) $\displaystyle\int \frac{x^2-1}{x\left(x^2+4\right)}\, dx = \frac{1}{4}\int\left(-\frac{1}{x} + \frac{5x}{x^2+4}\right) dx = -\frac{\ln|x|}{4} + \frac{5\ln\left|x^2+4\right|}{8} + C.$

21. Since $\displaystyle\left(\frac{4}{1-x} + 5\ln|3+x| + C\right)' = \frac{4}{(1-x)^2} + \frac{5}{3+x} = \frac{17 - 6x + 5x^2}{(1-x)^2(3+x)}$,

$q(x) = 17 - 6x + 5x^2.$

23. No, the antiderivative of a rational function never involves the arcsine function.

25. (a) After multiplying, the result is $x^2 = A(x+1)^2 + B(x+1) + C$. Substituting $x = -1$ into this equation yields $C = 1$.

(b) Since $C = 1$, the partial fraction decomposition equation is $\displaystyle\frac{x^2}{(x+1)^3} = \frac{A}{x+1} + \frac{B}{(x+1)^2} + \frac{1}{(x+1)^3}$. Substituting $x = 0$ into this equation yields $0 = A + B + 1$. Thus, $A + B = -1$.

(c) Substituting $C = 1$ and $x = 1$ into the partial fraction decomposition equation yields $\dfrac{1}{8} = \dfrac{A}{2} + \dfrac{B}{4} + \dfrac{1}{8}$. Thus, $2A + B = 0$.

(d) $A = 1$ and $B = -2$ is the unique solution of the system of equations $A + B = -1$ and $2A + B = 0$.

(e) $\displaystyle\int \frac{x^2}{(x+1)^3}\,dx = \int \left(\frac{1}{x+1} - \frac{2}{(x+1)^2} + \frac{1}{(x+1)^3} \right)\,dx$

$\displaystyle\qquad\qquad = \ln|x+1| + \frac{2}{x+1} - \frac{1}{2(x+1)^2} + C.$

27. (a) The partial fraction decomposition of $1/p(x)$ has the form $\dfrac{1}{p(x)} = \dfrac{A}{x+1} + \dfrac{B}{x-2}$.
    Solving this equation yields $A = -1/3$ and $B = 1/3$.

   (b) The partial fraction decomposition of $x/p(x)$ has the form $\dfrac{x}{p(x)} = \dfrac{A}{x+1} + \dfrac{B}{x-2}$.
    Solving this equation yields $A = 1/3$ and $B = 2/3$.

   (c) Parts (a) and (b) imply that $\dfrac{4-3x}{(x+1)(x-2)} = -\dfrac{2}{3(x-2)} - \dfrac{7}{3(x+1)}$. Thus,

$$\int \frac{4-3x}{(x+1)(x-2)}\,dx = -\frac{1}{3}\int \left(\frac{2}{x-2} + \frac{7}{x+1} \right)\,dx$$

$$= -\frac{2\ln|x-2|}{3} - \frac{7\ln|x+1|}{3} + C.$$

29. (a) The partial fraction decomposition of $1/p(x)$ has the form $\dfrac{1}{p(x)} = \dfrac{A}{x-1} + \dfrac{Bx+C}{x^2+1}$.
    Solving this equation yields $A = 1/2$, $B = -1/2$, and $C = -1/2$.

   (b) The partial fraction decomposition of $x/p(x)$ has the form $\dfrac{x}{p(x)} = \dfrac{A}{x-1} + \dfrac{Bx+C}{x^2+1}$.
    Solving this equation yields $A = 1/2$, $B = -1/2$, and $C = 1/2$.

   (c) The partial fraction decomposition of $x^2/p(x)$ has the form $\dfrac{x^2}{p(x)} = \dfrac{A}{x-1} + \dfrac{Bx+C}{x^2+1}$.
    Solving this equation yields $A = 1/2$, $B = 1/2$, and $C = 1/2$.

   (d) Parts (a)–(c) imply that $\dfrac{2+3x+4x^2}{(x-1)(x^2+1)} = \dfrac{9}{2(x-1)} + \dfrac{5-x}{2(x^2+1)}$. Thus,

$$\int \frac{2+3x+4x^2}{(x-1)(x^2+1)}\,dx = \frac{9\ln|x-1|}{2} + \frac{5\arctan x}{2} - \frac{\ln(x^2+1)}{4} + C.$$

31. The partial fraction decomposition of the integrand has the form

$$\frac{1}{(x-2)(x^2+1)} = \frac{A}{x-2} + \frac{Bx+C}{x^2+1}.$$

To determine the constants, multiply both sides of the equation above by $(x-2)(x^2+1)$ to obtain

$$1 = A(x^2+1) + (Bx+C)(x-2) = (A+B)x^2 + (C-2B)x + (A-2C).$$

It follows that $A + B = 0$, $C - 2B = 0$, and $A - 2C = 1$. Inserting $x = 2$ into the equation displayed above, we find that $A = 1/5$ and, therefore, $B = -1/5$ and $C = -2/5$. Thus,

$$\int_0^1 \frac{dx}{(x-2)(x^2+1)} = \frac{1}{5}\int_0^1 \frac{dx}{x-2} - \frac{1}{5}\int_0^1 \frac{x}{x^2+1}\,dx - \frac{2}{5}\int_0^1 \frac{dx}{x^2+1}$$

$$= \left(\frac{1}{5}\ln|x-2| - \frac{1}{10}\ln|x^2+1| - \frac{2}{5}\arctan x\right)\bigg]_0^1$$

$$= \left(-\frac{1}{10}\ln 2 - \frac{2}{5}\frac{\pi}{4}\right) - \frac{1}{5}\ln 2 = -\frac{3}{10}\ln 2 - \frac{\pi}{10}.$$

33. $\displaystyle\int \frac{2x+1}{(x-2)(x+3)}\,dx = \int\left(\frac{1}{x-2} + \frac{1}{x+3}\right)dx = \ln|x-2| + \ln|x+3| + C.$

35. $\displaystyle\int \frac{5x^2+3x-2}{x^3+2x^2}\,dx = \int\left(-\frac{1}{x^2} + \frac{2}{x} + \frac{3}{x+2}\right)dx = \frac{1}{x} + 2\ln|x| + 3\ln|x+2| + C.$

37. $\displaystyle\int \frac{x^4}{x^4-1}\,dx = \int\left(1 + \frac{1}{4}\cdot\frac{1}{x-1} - \frac{1}{4}\cdot\frac{1}{x+1} - \frac{1}{2}\cdot\frac{1}{1+x^2}\right)dx$

$$= x + \frac{1}{4}\ln|x-1| - \frac{1}{4}\ln|x+1| - \frac{1}{2}\arctan x + C.$$

39. $\displaystyle\int \frac{x^3}{x^2-1}\,dx = \int\left(x + \frac{1}{2}\cdot\frac{1}{x-1} + \frac{1}{2}\cdot\frac{1}{x+1}\right)dx$

$$= \frac{1}{2}x^2 + \frac{1}{2}\ln|x-1| + \frac{1}{2}\ln|x+1| + C.$$

41. $\displaystyle\int \frac{dx}{x\sqrt{x+1}} = 2\int\frac{du}{u^2-1} = \int\left(\frac{1}{u-1} - \frac{1}{u+1}\right)du$

$$= \ln\left|\frac{u-1}{u+1}\right| + C = \ln\left|\frac{\sqrt{x+1}-1}{\sqrt{x+1}+1}\right| + C.$$

43. (a) Since $\left(\ln|x+c|\right)' = \dfrac{1}{x+c}$, $\displaystyle\int\frac{dx}{x+c} = \ln|x+c|.$

(b) Since $\left(\dfrac{1}{1-n}\dfrac{1}{(x+c)^{n-1}}\right)' = \dfrac{1}{(x+c)^n}$, $\displaystyle\int\frac{dx}{(x+c)^n} = \frac{1}{(1-n)(x+c)^{n-1}}.$

(c) Using the substitution $u = ax$ in parts (a) and (b) (and replacing the constant c by b) leads to the desired result.

45. Let $u = x^2 + d^2$. Then, $\displaystyle\int \frac{x}{x^2+d^2}\,dx = \frac{1}{2}\int\frac{du}{u} = \frac{1}{2}\ln|u| = \frac{1}{2}\ln\left(x^2+d^2\right).$

47. Let $u = x^2 + d^2$. Then,

$$\int \frac{x}{\left(x^2+d^2\right)^n}\,dx = \frac{1}{2}\int\frac{du}{u^n} = \frac{1}{2(1-n)u^{n-1}} = \frac{1}{2(1-n)\left(x^2+d^2\right)^{n-1}}.$$

§8.3 Trigonometric Antiderivatives

1. (a) $\int \cos x \, dx = \sin x + C.$

   (b) $\int \sin x \, dx = -\cos x + C.$

   (c) $\int \sec^2 x \, dx = \tan x + C.$

   (d) $\int \sec x \tan x \, dx = \sec x + C.$

3. No—the two answers are equal. To see this, use the identity $\cos^2 x = 1 - \sin^2 x.$

5. (a) $\int \cos^3 x \sin^4 x \, dx = \int (1 - u^2)u^4 \, du = u^5/5 - u^7/7 = \dfrac{\sin^5 x}{5} - \dfrac{\sin^7 x}{7} + C.$

   (b) $\int \cos^3 x \sin^4 x \, dx = \int \cos^3 x \left(1 - \cos^2 x\right)^2 \, dx = \int \left(\cos^3 x - 2\cos^5 x + \cos^7 x\right) dx$

   $$= \int \cos^3 x \, dx - 2\int \cos^5 x \, dx + \frac{\cos^6 x \sin x}{7} + \frac{6}{7}\int \cos^5 x \, dx$$

   $$= \int \cos^3 x \, dx - \frac{8\cos^4 x \sin x}{35} - \frac{32}{35}\int \cos^3 x \, dx + \frac{\cos^6 x \sin x}{7}$$

   $$= \frac{\cos^2 x \sin x}{35} + \frac{2}{35}\int \cos x \, dx - \frac{8\cos^4 x \sin x}{35} + \frac{\cos^6 x \sin x}{7}$$

   $$= \frac{\cos^2 x \sin x}{35} + \frac{2\sin x}{35} - \frac{8\cos^4 x \sin x}{35} + \frac{\cos^6 x \sin x}{7} + C.$$

   (c) Yes, trigonometric identities can be used to show that the two expressions define the same function.

7. (a) $\cos t = \sqrt{a^2 - x^2}/a$ [adjacent/hypotenuse].

   (b) $\tan t = x/\sqrt{a^2 - x^2}$ [opposite/adjacent].

   (c) $\sin(2t) = 2\sin t \cos t = 2x\sqrt{a^2 - x^2}/a^2.$

9. $\displaystyle\int \frac{dx}{\sqrt{x^2 + 2x + 5}} = \int \frac{du}{\sqrt{u^2 + 4}} = \ln\left|\sqrt{u^2 + 4} + u\right|$

   $$= \ln\left|\sqrt{x^2 + 2x + 5} + x + 1\right| + C.$$

11. $\displaystyle\int \sin^2(3x) \, dx = \frac{1}{2}\int \left(1 - \cos(6x)\right) dx = \frac{1}{2}x - \frac{1}{12}\sin(6x) + C.$

13. Let $u = \sin x.$ Then, $\displaystyle\int \sin^2 x \cos x \, dx = \int u^2 \, du = u^3/3 = \sin^3 x/3 + C.$

15. $\displaystyle\int \cos^2 x \sin^3 x \, dx = \int \left(\cos^2 x - \cos^4 x\right) \sin x \, dx = -\frac{1}{3}\cos^3 x + \frac{1}{5}\cos^5 x + C.$

17. $\int \sin^2 x \cos^2 x \, dx = \int \left(\sin^2 x - \sin^4 x\right) dx = \frac{1}{8}x + \frac{1}{8}\cos x \sin x - \frac{1}{4}\cos^3 x \sin x + C.$

19. $\int \tan^4 x \, dx = \frac{1}{3}\tan^3 x - \tan x + x + C.$

21. $\int \sec^2 x \tan^2 x \, dx = \frac{1}{3}\tan^3 x + C.$

23. $\int \sec x \tan^2 x \, dx = \int \sec x (\sec^2 x - 1) \, dx = \int \sec^3 x \, dx - \int \sec x \, dx$

$$= \frac{1}{2}\left(\sec x \tan x + \int \sec x \, dx\right) - \int \sec x \, dx$$

$$= \frac{1}{2}\left(\tan x \sec x - \ln|\sec x + \tan x|\right) + C.$$

25. $\int \sec x \tan^3 x \, dx = \int \sec x \tan x \tan^2 x \, dx = \int \sec^2 x \sec x \tan x \, dx - \int \sec x \tan x \, dx$

$$= \sec^3 x/3 - \sec x + C.$$

27. $\int \frac{dx}{x^2\sqrt{4-x^2}} = \frac{1}{4}\int \csc^2 t \, dt = -\frac{1}{4}\cot t + C = -\frac{1}{4}\frac{\cos t}{\sin t} + C = -\frac{\sqrt{4-x^2}}{4x} + C$

$[x = 2\sin t, dx = 2\cos t \, dt, \cos t = \sqrt{1 - x^2/4}].$

29. $\int \sqrt{1-x^2}\, dx = \int \sqrt{1-\sin^2 t}\,\cos t \, dt = \int \cos^2 t \, dt = \frac{1}{2}\cos t \sin t + \frac{1}{2}t + C$

$$= \frac{1}{2}x\sqrt{1-x^2} + \frac{1}{2}\arcsin x + C.$$

31. $\int x^2\sqrt{1-x^2}\, dx = \int \sin^2 t \cos^2 t \, dt = \int \left(\cos^2 t - \cos^4 t\right) dt$

$$= \frac{t}{8} + \frac{1}{16}\sin(2t) - \frac{1}{4}\cos^3 t \sin t + C$$

$$= \frac{1}{8}\arcsin x + \frac{1}{8}x\sqrt{1-x^2} - \frac{1}{4}x\left(1-x^2\right)^{3/2} + C$$

$[x = \sin t, dx = \cos t \, dt, \cos t = \sqrt{1-x^2}].$

33.   (a) $\cos t = 1/\sec t = a/x$ [adjacent/hypotenuse].

(b) $\sin t = \sqrt{x^2 - a^2}/x$ [opposite/hypotenuse].

(c) $\tan(2t) = \dfrac{2\tan t}{1 - \tan^2 t} = \dfrac{2\tan t}{2 - \sec^2 t} = \dfrac{2a\sqrt{x^2 - a^2}}{2a^2 - x^2}.$

35. $\int \frac{dx}{x^2\sqrt{x^2-4}} = \frac{1}{4}\int \cos t \, dt = \frac{1}{4}\sin t + C = \frac{\sqrt{x^2-4}}{4x} + C$

$[x = 2\sec t, dx = 2\sec t \tan t \, dt, \sin t = \sqrt{1 - 4/x^2}].$

37. Let $x = \sec t$. When $x > 0$, $\sqrt{x^2 - 1} = \tan t$ (since $\tan t > 0$). Thus,

$$\int_1^2 \frac{\sqrt{x^2 - 1}}{x}\, dx = \int_0^{\pi/3} \tan^2 t\, dt = \tan t - t\Big]_0^{\pi/3} = \sqrt{3} - \pi/3.$$

39. $\displaystyle \int \frac{\sin^3 x}{\cos x}\, dx = \int \frac{(1 - \cos^2 x)\sin x}{\cos x}\, dx = \frac{1}{2}\cos^2 x - \ln|\cos x| + C.$

41. $\displaystyle \int \sqrt{\cos x}\, \sin^5 x\, dx = \int \sqrt{\cos x}\, (1 - \cos^2 x)^2 \sin x\, dx$

$$= -\frac{2}{3}(\cos x)^{3/2} + \frac{4}{7}(\cos x)^{7/2} - \frac{2}{11}(\cos x)^{11/2} + C.$$

43. $\displaystyle \int \sqrt{1 + x^2}\, dx = \int \sec t \sec^2 t\, dt = \int \sec^3 t\, dt = \frac{1}{2}\tan t \sec t + \frac{1}{2}\ln|\sec t + \tan t| + C$

$$= \frac{1}{2}x\sqrt{1 + x^2} + \frac{1}{2}\ln\left|\sqrt{1 + x^2} + x\right| + C$$

$[x = \tan t,\ dx = \sec^2 t\, dt,\ \sec t = \sqrt{1 + x^2}].$

45. First, note that
$$\int \frac{x + 2}{x(x^2 + 1)}\, dx = \int \frac{dx}{x^2 + 1} + 2\int \frac{dx}{x(x^2 + 1)} = \arctan x + 2\int \frac{dx}{x(x^2 + 1)}.$$ Also,

$$\int \frac{dx}{x(x^2 + 1)} = \int \frac{\cos t}{\sin t}\, dt = \ln|\sin t| + C = \ln\left|\frac{x}{\sqrt{1 + x^2}}\right| + C.$$

$[x = \tan t,\ dx = \sec^2 t\, dt,\ \sin t = x/\sqrt{1 + x^2}].$

Therefore, $\displaystyle \int \frac{x + 2}{x(x^2 + 1)}\, dx = \arctan x + 2\ln\left|\frac{x}{\sqrt{1 + x^2}}\right| + C.$

47. $\displaystyle \int \frac{\arctan x}{(1 + x^2)^{3/2}}\, dx = \int w \cos w\, dw = w \sin w + \cos w = \frac{1 + x \arctan x}{\sqrt{1 + x^2}} + C.$

49. (a) Since $\cos(-v) = \cos v$ and $\sin(-v) = -\sin v$, $\cos(u - v) = \cos u \cos v + \sin u \sin v$.
    Therefore, $\cos(u + v) + \cos(u - v) = (\cos u \cos v - \sin u \sin v) +$
    $(\cos u \cos v + \sin u \sin v) = 2\cos u \cos v.$

    (b) $\displaystyle \int \cos(ax)\cos(bx)\, dx = \frac{1}{2}\int \Big(\cos((a + b)x) + \cos((a - b)x)\Big)\, dx$

    $$= \frac{1}{2(a + b)}\sin((a + b)x) + \frac{1}{2(a - b)}\sin((a - b)x) + C$$

51. (a) Since $\cos(-v) = \cos v$ and $\sin(-v) = -\sin v$, $\sin(u - v) = \sin u \cos v - \cos u \sin v$.
    Therefore,
    $\sin(u + v) + \sin(u - v) = (\sin u \cos v + \cos u \sin v) + (\sin u \cos v - \cos u \sin v)$
    $= 2\sin u \cos v.$

(b) $\displaystyle\int \sin(ax)\cos(bx)\,dx = \frac{1}{2}\int \Big(\sin\big((a+b)x\big)+\sin\big((a-b)x\big)\Big)\,dx$

$$= -\frac{1}{2(a+b)}\cos\big((a+b)x\big) - \frac{1}{2(a-b)}\cos\big((a-b)x\big) + C$$

53. (a) $u=\sin^{n-1}x \implies du=(n-1)\sin^{n-2}x\cos x\,dx;\ dv=\sin x\,dx \implies v=-\cos x.$
    Thus,

$$\int \sin^n x\,dx = \int u\,dv = uv - \int v\,du = -\sin^{n-1}x\cos x + (n-1)\int \sin^{n-2}x\cos^2 x\,dx.$$

(b) $\displaystyle\int \sin^{n-2}x\cos^2 x\,dx = \int \sin^{n-2}x\left(1-\sin^2 x\right)dx = \int \sin^{n-2}x\,dx - \int \sin^n x\,dx.$
    Thus,

$$\int \sin^n x\,dx = -\sin^{n-1}x\cos x + (n-1)\int \sin^{n-2}x\,dx - (n-1)\int \sin^n x\,dx$$

so $\displaystyle n\int \sin^n x\,dx = -\sin^{n-1}x\cos x + (n-1)\int \sin^{n-2}x\,dx.$ Therefore, if $n \neq 0$,

$$\int \sin^n x\,dx = -\frac{\sin^{n-1}x\cos x}{n} + \frac{n-1}{n}\int \sin^{n-2}x\,dx.$$

55. Let $\displaystyle I_n = \int_0^{\pi/2}\sin^n x\,dx$. Then $I_1 = 1$, $I_2 = \pi/4$, and the reduction formula implies that $I_n = \frac{n-1}{n}I_{n-2}$.

If n is an even integer such that $n > 2$,

$$I_n = \frac{n-1}{n}I_{n-2} = \frac{n-1}{n}\frac{n-3}{n-2}I_{n-4} = \cdots = \frac{(n-1)}{n}\frac{(n-3)}{n-2}\cdots\frac{5}{6}\frac{3}{4}\frac{\pi}{4}.$$

57. $\displaystyle\int \tan^n x\,dx = \int \tan^{n-2}x\tan^2 x\,dx = \int \tan^{n-2}x(\sec^2 x - 1)\,dx$

$$= \int \tan^{n-2}x\sec^2 x\,dx - \int \tan^{n-2}x\,dx = \frac{\tan^{n-1}x}{n-1} - \int \tan^{n-2}x\,dx.$$

59. Draw a right triangle with hypotenuse $\sqrt{1+t^2}$ and sides of length 1 and t.

    (a) If the angle opposite the side of length t is $x/2$, then $\sin(x/2) = t/\sqrt{1+t^2}$ since $\sin\theta = $ opposite/hypotenuse.

    (b) If the angle opposite the side of length t is $x/2$, then $\cos(x/2) = 1/\sqrt{1+t^2}$ since $\cos\theta = $ adjacent/hypotenuse.

    (c) The desired result follows from the identity $\sin(2\theta) = 2\cos\theta\sin\theta$.

(d) The desired result follows from the identity $\cos(2\theta) = 2\cos^2\theta - 1$.

61. $\displaystyle\int \frac{dx}{1+\cos x} = \int \frac{1}{1+(1-t^2)/(1+t^2)} \frac{2}{1+t^2}\,dt = \int dt = t + C = \tan(x/2) + C.$

63. Using the substitution $u = 2 + \cos x$, $\displaystyle\int \frac{\sin x}{2+\cos x}\,dx = -\int \frac{du}{u} = -\ln(2+\cos x)$. Thus,

$$\int_{\pi/4}^{\pi} \frac{\sin x}{2+\cos x}\,dx = -\ln(2+\cos x)\Big]_{\pi/4}^{\pi} = \ln(2+\sqrt{2}/2) = \ln(4+\sqrt{2}) - \ln 2.$$

65. (a) Let $u = 1/(1+x^2)^n$ and $dv = dx$. Then, $du = -2nx/(1+x^2)^{n+1}\,dy$ and $v = x$ so

$$\begin{aligned}
I_n = \int \frac{dx}{(1+x^2)^n} &= \frac{x}{(1+x^2)^n} + 2n\int \frac{x^2}{(1+x^2)^{n+1}}\,dx \\
&= \frac{x}{(1+x^2)^n} + 2n\int \frac{1+x^2}{(1+x^2)^{n+1}}\,dx - 2n\int \frac{dx}{(1+x^2)^{n+1}} \\
&= \frac{x}{(1+x^2)^n} + 2n\int \frac{dx}{(1+x^2)^n} - 2n\int \frac{dx}{(1+x^2)^{n+1}}.
\end{aligned}$$

Thus,

$$I_{n+1} = \frac{x}{2n(1+x^2)^n} - \frac{1-2n}{2n}I_n$$

or, equivalently,

$$I_n = \frac{x}{2(n-1)(1+x^2)^{n-1}} + \frac{2n-3}{2n-2}I_{n-1}.$$

(b) If $x = \tan u$, then $dx = \sec^2 u\,du$ and $I_n = \displaystyle\int \frac{dx}{(1+x^2)^n} = \int \cos^{2n-2} u\,du$. Now, using the reduction formula for powers of the cosine function,

$$\begin{aligned}
I_n = \int \frac{dx}{(1+x^2)^n} &= \int \cos^{2n-2} u\,du \\
&= \frac{\cos^{2n-3} u \sin u}{2n-2} + \frac{2n-3}{2n-2}\int \cos^{2n-4} u\,du \\
&= \frac{x}{(2n-2)(1+x^2)^{n-1}} + \frac{2n-3}{2n-2}I_{n-1}.
\end{aligned}$$

§8.4 Miscellaneous Antiderivatives

1. $\displaystyle\int \frac{\sin x}{(3+\cos x)^2}\,dx = \frac{1}{3+\cos x} + C.$

   [substitution: $u = 3 + \cos x.$]

3. $\displaystyle\int x\left(3+4x^2\right)^5\,dx = \frac{1}{48}\left(3+4x^2\right)^6 + C.$

   [substitution: $u = 3 + 4x^2.$]

5. $\displaystyle\int \frac{x}{\sqrt[3]{x^2+4}}\,dx = \frac{3}{4}\left(x^2+4\right)^{2/3} + C.$

   [substitution: $u = x^2 + 4.$]

7. $\displaystyle\int \frac{(\ln x)^2}{x}\,dx = \frac{1}{3}\left(\ln|x|\right)^3 + C.$

   [substitution: $u = \ln x.$]

9. $\displaystyle\int \frac{\ln x}{x}\,dx = \frac{1}{2}\left(\ln|x|\right)^2 + C.$

   [substitution: $u = \ln x$] .

11. $\displaystyle\int \frac{x}{3x+2}\,dx = \frac{1}{3}\int\left(1 - \frac{2}{3x+2}\right)dx = \frac{x}{3} - \frac{2}{9}\ln|3x+2| + C.$

13. $\displaystyle\int \sin^2(3x)\cos(3x)\,dx = \frac{1}{9}\sin^3(3x) + C.$

    [substitution: $u = \sin(3x).$]

15. $\displaystyle\int xe^{3x^2}\,dx = \frac{1}{6}e^{3x^2} + C.$

    [substitution: $u = 3x^2.$]

17. $\displaystyle\int (2-3x)^{10}\,dx = -\frac{1}{33}(2-3x)^{11} + C.$

    [substitution: $u = 2 - 3x.$]

19. $\displaystyle\int \frac{\sec^2 x}{3+\tan x}\,dx = \ln|3+\tan x| + C.$

    [substitution: $u = 3 + \tan x.$]

21. $\displaystyle\int \frac{dx}{(x-1)(x+2)} = \frac{1}{3}\ln|x-1| - \frac{1}{3}\ln|x+2| = \frac{1}{3}\ln\left|\frac{x-1}{x+2}\right| + C.$

    [partial fractions: $\frac{1}{(x-1)(x+2)} = \frac{1}{3(x-1)} - \frac{1}{3(x+2)}.$]

23. $\displaystyle\int \frac{2x+3}{4x+5}\,dx = \frac{1}{2}\int\left(1 + \frac{1}{4x+5}\right)dx = \frac{x}{2} + \frac{\ln|4x+5|}{8} + C.$

$\dfrac{1}{16}\left(x + \ln|4x+5|\right) + c.$

25. $\int \dfrac{e^x}{\sqrt{1-e^{2x}}}\, dx = \arcsin\left(e^x\right) + C.$

    [substitution: $u = e^x$.]

27. $\int \ln x\, dx = x(\ln x - 1) + C.$

    [integration by parts: $u = \ln x,\, dv = dx$.]

29. $\int \arcsin x\, dx = x \arcsin x + \sqrt{1 - x^2} + C.$

    [integration by parts: $u = \arcsin x,\, dv = dx$.]

31. $\int \dfrac{dx}{x^2 + 2x + 3} = \dfrac{1}{\sqrt{2}} \arctan\left(\dfrac{x+1}{\sqrt{2}}\right) + C.$

    [complete the square, substitution: $x^2 + 2x + 3 = (x+1)^2 + 2 = u^2 + 2$.]

33. $\int \dfrac{dx}{\sqrt{1 - 4x^2}} = \dfrac{1}{2} \arcsin(2x) + C.$

    [substitution: $u = 2x$.]

35. $\int \tan x\, dx = -\ln|\cos x| + C.$

    [Write $\tan x = \sin x / \cos x$, then use the substitution $u = \cos x$.]

37. $\int e^{2x}\sqrt{1 + e^x}\, dx = \dfrac{2}{5}\left(1 + e^x\right)^{5/2} - \dfrac{2}{3}\left(1 + e^x\right)^{3/2} + C.$

    [substitution: $u = 1 + e^x$.]

39. Let $u = (x + 1)/2$. Then, $du = \frac{1}{2}dx$ and
    $$\int \dfrac{dx}{\sqrt{3 - 2x - x^2}} = \int \dfrac{du}{\sqrt{1 - u^2}} = \arcsin u = \arcsin\left(\tfrac{1}{2}(x+1)\right) + C.$$

41. Let $u = \ln x$. Then, $du = dx/x$ and $\int \dfrac{dx}{x(\ln x)^2} = \int \dfrac{du}{u^2} = -\dfrac{1}{u} + C = -\dfrac{1}{\ln|x|} + C.$

43. $\int \dfrac{dx}{9x^2 - 4} = \int \dfrac{dx}{(3x - 2)(3x + 2)} = \dfrac{1}{4}\int\left(\dfrac{1}{3x - 2} - \dfrac{1}{3x + 2}\right)dx$
    $$= \left(\ln|3x - 2| - \ln|3x + 2|\right)/12 + C.$$

45. $\int \dfrac{x^3}{\sqrt{4 - x^2}}\, dx = -4\sqrt{4 - x^2} + \dfrac{1}{3}\left(4 - x^2\right)^{3/2} + C.$

    [substitution $(u = 4 - x^2)$ or integration by parts $(u = x^2,\, dv = x/\sqrt{4 - x^2}\, dx)$.]

47. $\int \dfrac{x}{(x - 1)(x + 1)}\, dx = \dfrac{1}{2}\ln|x + 1| + \dfrac{1}{2}\ln|x - 1| = \dfrac{1}{2}\ln\left|x^2 - 1\right| + C.$

    [partial fractions: $\frac{x}{(x-1)(x+1)} = \frac{1}{2(x+1)} + \frac{1}{2(x-1)}$.]

49. $\displaystyle\int \frac{dx}{\sqrt{9+x^2}} = \ln\left|x + \sqrt{9+x^2}\right| + C.$

[trigonometric substitution: $x = 3\tan t$.]

51. $\displaystyle\int \frac{x^2}{1-3x}\,dx = -\frac{1}{54}(1-3x)^2 + \frac{2}{27}(1-3x) - \frac{1}{27}\ln|1-3x| + C$

$\qquad\qquad = -\frac{1}{6}x^2 - \frac{1}{9}x - \frac{1}{27}\ln|1-3x| + C.$

[substitution ($u = 1 - 3x$) or partial fractions.]

53. $\displaystyle\int e^x e^{2x}\,dx = \int e^{3x}\,dx = \frac{1}{3}e^{3x} + C.$

55. $\displaystyle\int \ln(1+x^2)\,dx = x\ln(1+x^2) - 2x + 2\arctan x + C.$

[integration by parts: $u = \ln(1+x^2), dv = dx$.]

57. Using integration by parts $\displaystyle\int x\arcsin x\,dx = \frac{x^2\arcsin x}{2} - \frac{1}{2}\int \frac{x^2}{\sqrt{1-x^2}}\,dx.$ Now, using

the trigonometric substitution $x = \sin t$,

$$\int \frac{x^2}{\sqrt{1-x^2}}\,dx = \int \sin^2 t\,dt = (-\sin t\cos t + t)/2$$

$$= (-x\sqrt{1-x^2} + \arcsin x)/2.$$

Thus,

$$\int x\arcsin x\,dx = \frac{x^2\arcsin x}{2} + \frac{x\sqrt{1-x^2}}{4} - \frac{\arcsin x}{4} + C.$$

59. Let $u = 2x + 3$. Then, $\displaystyle\int \frac{dx}{\sqrt{2x+3}} = \frac{1}{2}\int \frac{du}{\sqrt{u}} = \sqrt{u} = \sqrt{2x+3} + C.$

61. Let $u = 2x + 3$. Then,

$$\int \frac{x}{(2x+3)^4}\,dx = \frac{1}{4}\int \frac{u-3}{u^4}\,du = -\frac{1}{8u^2} + \frac{1}{4u^3}$$

$$= -\frac{1}{8(2x+3)^2} + \frac{1}{4(2x+3)^3} = -\frac{2x+1}{8(2x+3)^3} + C.$$

63. $\displaystyle\int \frac{\tan x}{\sec^2 x}\,dx = -\frac{1}{2}\cos^2 x + C.$

[Write $\frac{\tan x}{\sec^2 x} = \sin x\cos x$, then use substitution ($u = \cos x$).]

65. $\displaystyle\int \frac{dx}{e^x - 1} = \ln\left|1 - e^{-x}\right| + C.$

[Write $\frac{1}{e^x-1} = \frac{e^{-x}}{1-e^{-x}}$, then use subsitution $u = 1 - e^{-x}$.]

67. $\int \dfrac{dx}{1 + \sqrt{x}} = 2(1 + \sqrt{x}) - 2\ln(1 + \sqrt{x}) + C.$

    [subsitution: $u = 1 + \sqrt{x}$.]

69. $\int x^2 \ln(3x)\, dx = \dfrac{1}{3} x^3 \ln(3x) - \dfrac{1}{9} x^3 + C.$

    [integration by parts: $u = \ln(3x),\, dv = x^2\, dx.$]

71. $\int \sqrt{x}\, \ln x\, dx = \dfrac{2}{3} x^{3/2} \ln|x| - \dfrac{4}{9} x^{3/2} + C.$

    [integration by parts: $u = \ln x,\, dv = \sqrt{x}\, dx.$]

73. $\int \dfrac{7 - x}{(x + 3)(x^2 + 1)}\, dx = \ln|x + 3| + 2\arctan x - \dfrac{1}{2} \ln(x^2 + 1) + C.$

    [partial fractions: $\frac{7-x}{(x+3)(x^2+1)} = \frac{1}{x+3} + \frac{2-x}{x^2+1}.$]

75. $\int x \sin^2 x \cos x\, dx = \dfrac{1}{3} x \sin^3 x + \dfrac{1}{3} \cos x - \dfrac{1}{9} \cos^3 x$

    $\qquad\qquad = \dfrac{1}{3} x \sin^3 x + \dfrac{1}{9} \sin^2 x \cos x + \dfrac{2}{9} \cos x + C.$

    [integration by parts with $u = x$ and $dv = \sin^2 x \cos x\, dx.$]

77. $\int \dfrac{dx}{x^3 + x} = \ln|x| - \dfrac{1}{2} \ln(x^2 + 1) + C.$

    [partial fractions: $\frac{1}{x^3+x} = \frac{1}{x} - \frac{x}{x^2+1}.$]

79. $\int \left(x^2 + 2x + 3\right)^{3/2} dx = \dfrac{1}{4}(x + 1)(x^2 + 2x + 3)^{3/2} + \dfrac{3}{4}(x + 1)\sqrt{x^2 + 2x + 3} +$

    $\dfrac{3}{2} \ln \left| \dfrac{\sqrt{x^2 + 2x + 3}}{\sqrt{2}} + \dfrac{x + 1}{\sqrt{2}} \right| + C.$

    [Write $x^2 + 2x + 3 = (x + 1)^2 + 2$, then use a trigonometric substitution $(x + 1 = \sqrt{2} \tan t)$.]

81. Let $u = \sqrt{1 + e^x}$. Then,

$$\int \sqrt{1 + e^x}\, dx = 2 \int \dfrac{u^2}{u^2 - 1}\, du = 2 \int \left(1 + \dfrac{1}{u^2 - 1}\right) du$$

$$= 2 \int \left(1 - \dfrac{1}{2(u + 1)} + \dfrac{1}{2(u - 1)}\right) du$$

$$= 2u - \ln|u + 1| + \ln|u - 1|$$

$$= 2\sqrt{1 + e^x} - \ln\left|\sqrt{1 + e^x} + 1\right| + \ln\left|\sqrt{1 + e^x} - 1\right| + C.$$

83. $\int \dfrac{dx}{x^3 + 1} = \dfrac{1}{3} \ln|x + 1| - \dfrac{1}{6} \ln|x^2 - x + 1| + \dfrac{1}{\sqrt{3}} \arctan\left(\dfrac{2x - 1}{\sqrt{3}}\right) + C.$

[Write $x^3 + 1 = (x + 1)(x^2 - x + 1) = (x + 1) \cdot \left((2x - 1)^2 + 3\right) / 4$, then use partial fractions, etc.]

85. $\displaystyle\int x \tan^2 x \, dx = x \tan x - \frac{1}{2}x^2 - \ln|\sec x| + C.$

[integration by parts: $u = x, \, dv = \tan^2 x \, dx.$]

87. $\displaystyle\int \sin x \sin(2x) \, dx = \frac{1}{2} \sin x - \frac{1}{6} \sin(3x) + C.$

[Write $\sin x \sin(2x) = \frac{1}{2} \cos x - \frac{1}{2} \cos(3x).$]

9 Function Approximation

§9.1 Taylor Polynomials

1. (a) If f is a polynomial of degree n, the Maclaurin series for f is just f itself:
   $1 + 2x + 44x^2 - 12x^3 + x^4$.

   (b) Finding values and derivatives at $x = 3$ gives
   $f(x) = 160 + 50(x - 3) - 10(x - 3)^2 + (x - 3)^4$.

3. Since p is a polynomial, it's easiest to find $p(1)$ and $p^{(4)}(1)$ from the coefficients of p:
   $p(1) = 9$ and $p^{(4)}(1)/24 = 1/5$.

   (a) $p(1) = 9$

   (b) $p^{(4)}(1) = 24/5$

5. Let a_k be the coefficient of $(x + 3)^k$. Then $a_k = p^{(k)}(-3)/k!$. In particular:

   (a) $p^{(11)}(-3) = a_{11} \times 11! = 0$.

   (b) $p^{(12)}(-3) = a_{12} \times 12! = \dfrac{1}{13^4} \times 12! \approx 16771$.

7. Finding values and derivatives at $x = 0$ gives $P_6(x) = 1 - \dfrac{x^2}{2} + \dfrac{x^4}{24} - \dfrac{x^6}{720}$; plotting f and P_6 together near $x = 0$ shows that the approximation error is less than 0.01 on $[-2.13, 2.13]$ (or on any smaller interval centered at zero).

9. Finding values and derivatives at $x = 0$ gives $P_6(x) = x - \dfrac{x^2}{2} + \dfrac{x^3}{3} - \dfrac{x^4}{4} + \dfrac{x^5}{5} - \dfrac{x^6}{6}$; plotting f and P_6 together near $x = 0$ shows that the approximation error is less than 0.01 on $[-.61, .73]$ (or on any smaller interval centered at zero).

11. Finding values and derivatives at $x = 0$ gives $P_6(x) = x + \dfrac{x^3}{6} + \dfrac{x^5}{120}$; plotting f and P_6 together near $x = 0$ shows that the approximation error is less than 0.01 on $[-1.74, 1.74]$ (or on any smaller interval centered at zero).

13. Finding values and derivatives at $x = \pi/4$ gives
   $$P_5(x) = \frac{1}{\sqrt{2}} \left(1 + x - \pi/4 - \frac{(x - \pi/4)^2}{2} - \frac{(x - \pi/4)^3}{6} + \frac{(x - \pi/4)^4}{24} + \frac{(x - \pi/4)^5}{120} \right);$$
   plotting f and P_5 together near $x = \pi/4$ shows that the approximation error is less than 0.01 on $[-0.76, 2.22]$ (or on any smaller interval centered at $\pi/4$).

15. Finding values and derivatives at $x = \pi/3$ gives

$$P_5(x) = \frac{\sqrt{3}}{2} + \frac{x - \pi/3}{2} - \frac{\sqrt{3}}{4}(x - \pi/3)^2 - \frac{1}{12}(x - \pi/3)^3$$
$$+ \frac{1}{16\sqrt{3}}(x - \pi/3)^4 + \frac{1}{240}(x - \pi/3)^5;$$

plotting f and P_5 together near $x = \pi/3$ shows that the approximation error is less than 0.01 on $[-0.41, 2.45]$ (or on any smaller interval centered at $\pi/3$).

17. Finding values and derivatives at $x = 1$ gives
$P_5(x) = 1 - (x - 1) + (x - 1)^2 - (x - 1)^3 + (x - 1)^4 - (x - 1)^5$; plotting f and P_5 together near $x = 1$ shows that the approximation error is less than 0.01 on $[0.58, 1.49]$ (or on any smaller interval centered at $x = 1$).

19. Finding values and derivatives at $x = 1$ gives
$$P_5(x) = (x - 1) - \frac{(x - 1)^2}{2} + \frac{(x - 1)^3}{3} - \frac{(x - 1)^4}{4} + \frac{(x - 1)^5}{5};$$ plotting f and P_5 together near $x = 1$ shows that the approximation error is less than 0.01 on $[0.45, 1.67]$ (or on any smaller interval centered at $x = 1$).

21. Finding values and derivatives at $x = 1$ gives
$$P_5(x) = 1 + \frac{x - 1}{2} - \frac{(x - 1)^2}{8} + \frac{(x - 1)^3}{16} - \frac{5(x - 1)^4}{128} + \frac{7(x - 1)^5}{256};$$ plotting f and P_5 together near $x = 1$ shows that the approximation error is less than 0.01 on $[0.25, 1.97]$ (or on any smaller interval centered at $x = 1$).

23. Finding values and derivatives at $x = 1$ gives
$$P_5(x) = 1 - \frac{x - 1}{2} + 3\frac{(x - 1)^2}{8} - 5\frac{(x - 1)^3}{16} + 35\frac{(x - 1)^4}{128} - 63\frac{(x - 1)^5}{256};$$ plotting f and P_5 together near $x = 1$ shows that the approximation error is less than 0.01 on $[0.47, 1.64]$ (or on any smaller interval centered at $x = 1$).

25. No. $P_n(x)$ involves powers of $(x - x_0)^k$ with k up to but not exceeding n.

27. $P_2(x)$ is the sum of terms of $P_5(x)$ up through degree 2.

29. If f is odd, then f' is even, f'' is odd, f''' is even, etc. Thus all even-order derivatives of f are odd functions, and so have the value zero for input zero. This implies that all even-order Maclaurin polynomial terms are zero.

31. Note that $ln(x) = f(x - 1)$; this fact and Exercise 30 imply that the desired polynomial is
$P_4(x) = (x - 1) - (x - 1)^2/2 + (x - 1)^3/3 - (x - 1)^4/4$.

33. Note that $g(x) = f(x + 1) = f(x - (-1))$; this fact and Exercise 30 imply that the desired polynomial is $P_4(x) = 1 - x/2 + 3x^2/8 - 5x^3/16 + 35x^4/128$.

35. For $f(x) = \sin x$ we have $M_5(x) = x - \dfrac{x^3}{6} + \dfrac{x^5}{120}$. Setting $k = 1/2$ in the result of Exercise 34 gives the new 5th-order Maclaurin polynomial

$$P_5(x) = \frac{x}{2} - \frac{x^3}{6 \cdot 8} + \frac{x^5}{120 \cdot 32} = \frac{x}{2} - \frac{x^3}{48} + \frac{x^5}{3840}.$$

37. For $f(x) = e^x$ we have $M_5(x) = 1 + x + \dfrac{x^2}{2} + \dfrac{x^3}{6} + \dfrac{x^4}{24} + \dfrac{x^5}{120}$, so replacing x with $-x$ gives the new 5th-order Maclaurin polynomial $P_5(x) = 1 - x + \dfrac{x^2}{2} - \dfrac{x^3}{6} + \dfrac{x^4}{24} - \dfrac{x^5}{120}$.

39. For $g(x) = e^x$ we have $M_3(x) = 1 + x + \dfrac{x^2}{2} + \dfrac{x^3}{6}$; replacing x with x^2 gives

$P_6(x) = 1 + x^2 + \dfrac{x^4}{2} + \dfrac{x^6}{6}$, the sixth-order Maclaurin polynomial for $f(x) = e^{x^2}$.

41. For $g(x) = 1/(1+x)$ we have $M_3(x) = 1 - x + x^2 - x^3$; replacing x with x^2 gives
$P_6(x) = 1 - x^2 + x^4 - x^6$, the sixth-order Maclaurin polynomial for $f(x) = 1/(1+x^2)$.

43. (a) Differentiating both sides of $g(x) = xf(x)$ repeatedly (using the product rule) shows the pattern: $g'(x) = xf'(x) + f(x)$,
$g''(x) = xf''(x) + f'(x) + f'(x) = xf''(x) + 2f'(x)$, $g'''(x) = xf'''(x) + 3f''(x)$, and so on.

   (b) Part (a) implies that $xM_n(x)$ and $xf(x)$ have the same value and first $n+1$ derivatives at $x = 0$. This means that $xM_n(x)$ is indeed the Maclaurin polynomial of order $n+1$ for $xf(x)$.

45. By the sum rule for derivatives, $h^{(k)}(x_0) = f^{(k)}(x_0) + g^{(k)}(x_0)$ for all orders k; similarly, $r^{(k)}(x_0) = p^{(k)}(x_0) + q^{(k)}(x_0)$. Thus, the value and first n derivatives of h agree with those of r at $x = x_0$.

47. Note that $F(x)$ and $P_n(x)$ have the same value and derivatives through order n at x_0. It follows that $F'(x)$ and $P_n'(x)$ have the same value and derivatives through order $n-1$ at x_0. Thus $P_n'(x)$ is the desired Taylor polynomial.

49. Note that $\cos x$ has 6th-order Maclaurin polynomial $P_6(x) = 1 - \dfrac{x^2}{2} + \dfrac{x^4}{24} - \dfrac{x^6}{720}$. It follows from this and the hint that $\cos^2 x$ has 6th-order Maclaurin polynomial

$$Q_6(x) = \frac{1 + P_6(2x)}{2} = 1 - x^2 + \frac{x^4}{3} - \frac{2x^6}{45}.$$

51. Note that $\ln(1+x)$ has 6th-order Maclaurin polynomial

$P_6(x) = x - \dfrac{x^2}{2} + \dfrac{x^3}{3} - \dfrac{x^4}{4} + \dfrac{x^5}{5} - \dfrac{x^6}{6}$. Replacing x with $-x$ shows that $\ln(1-x)$ has

6th-order Maclaurin polynomial $Q_6(x) = -x - \dfrac{x^2}{2} - \dfrac{x^3}{3} - \dfrac{x^4}{4} - \dfrac{x^5}{5} - \dfrac{x^6}{6}$. Subtracting these polynomials gives $R_6(x) = P_6(x) - Q_6(x) = 2x + \dfrac{2x^3}{3} + \dfrac{2x^5}{5}$, the 6th-order Maclaurin polynomial for the original function.

53. Note that $g(x) = 1/(1 + x^2)$ has 5th-order Maclaurin polynomial $P_5(x) = 1 - x^2 + x^4$. Integrating this gives $Q_6(x) = x - \dfrac{x^3}{3} + \dfrac{x^5}{5}$, the 6th-order Maclaurin polynomial for arctan x.

§9.2   Taylor's Theorem: Accuracy Guarantees for Taylor Polynomials

1. (a) The graph shows that $|f(x) - P_1(x)|$ is largest at $x = 80$, where its value is about 0.056.

    (b) Yes; the result seen in (a) is considerably less than the bound 0.091.

3. (a) Note that if $f(x) = \sin x$, then $f^{(6)}(x) = -\sin x$, and so $\left|f^{(6)}(x)\right| \le 1$ for all x. Thus we can use $K_6 = 1$ in Theorem 2, which says that the approximation error $|f(x) - P_5(x)|$ doesn't exceed $\frac{1}{6!}|x|^6$ for x in $[-2, 2]$. This quantity is largest when $x = 2$; in that case $\frac{1}{6!}|x|^6 = \frac{2^6}{6!} = \frac{4}{45} \approx 0.089$.

    (b) Plotting $|f(x) - P_5(x)|$ for $-2 \le x \le 2$ shows that the maximum approximation error occurs at $x = \pm 2$, where $|f(x) - P_5(x)| \approx 0.024$.

5. (a) If $f(x) = \dfrac{1}{\sqrt{x}}$, then $f^{(5)}(x) = -\dfrac{945}{32\, x^{11/2}}$. For x in $[1/2, 3/2]$, $|f^{(5)}(x)| \le |f^{(5)}(1/2)| \approx 1336.5$. Thus we can use $K_5 = 1336.5$ in Theorem 2, which says that the approximation error $|f(x) - P_4(x)|$ doesn't exceed $\frac{1336.5}{5!}|x - 1|^5$ for x in $[1/2, 3/2]$. This quantity is largest when $|x - 1| = 1/2$; in that case $\dfrac{1336.5}{5!}(1/2)^5 \approx 0.348$.

    (b) Plotting $|f(x) - P_4(x)|$ for $1/2 \le x \le 3/2$ shows that the maximum approximation error occurs at $x = 1/2$, where $|f(x) - P_4(x)| \approx 0.0143$.

7. Since $f(x) = \sin x$, all derivatives of f are sines or cosines, which never exceed 1 in absolute value. Hence we can always use $K_{n+1} = 1$ in Theorem 2, which then says that $$|f(x) - M_n(x)| \le \frac{|x|^{n+1}}{(n+1)!}.$$ For any fixed x the right side goes to zero as $n \to \infty$, and so is less than 10^{-6} for large n.

9. (a) The facts that $f(0) = 0$, $f'(0) = 0$, and $f''(0) = 2$ (all are easy to check) imply that $P_2(x) = x^2$.

    (b) Here $f'''(x) = -8x^3 \cos(x^2) - 12x \sin(x^2)$; a graph shows that $|f'''(x)| < 2.5$ for x in $[0, 1/2]$. Now Theorem 2, with $K_3 = 2.5$, gives $\left|\sin(t^2) - t^2\right| \le 2.5 t^3 / 3!$.

    (c) The result of part (b) implies that

$$\left|\int_0^{1/2} \sin(x^2)\, dx - \frac{1}{24}\right| = \left|\int_0^{1/2} \left(\sin(x^2) - x^2\right) dx\right|$$
$$\le \int_0^{1/2} \left|\sin(x^2) - x^2\right| dx \le \int_0^{1/2} 2.5 x^3 / 6\, dx$$
$$= \frac{5}{768} \approx 0.0065.$$

(d) The given inequality implies that

$$\left| \int_0^{1/2} \sin(x^2)\,dx - \int_0^{1/2} x^2\,dx \right| \le \int_0^{1/2} \left| \sin(x^2) - x^2 \right|\,dx$$

$$\le \int_0^{1/2} 120 x^6 / 6!\,dx \approx 0.00019.$$

11. Theorem 2 implies that $|x - \sin x| \le \dfrac{K_3}{3!}|x|^3$, where K_3 is a constant. (In fact, we can use $K_3 = 1$, but all that matters here is that K_3 is a constant.) Thus,

$$\left| \frac{x - \sin x}{x^2} \right| \le \frac{K_3 |x|^3}{3!\,x^2} = \frac{K_3}{3!}|x|.$$

It follows that $\lim\limits_{x \to 0} \left| \dfrac{x - \sin x}{x^2} \right| = 0$ and so $\lim\limits_{x \to 0} \dfrac{x - \sin x}{x^2} = 0.$

13. Theorem 2 implies that $\left| \cos x - 1 + x^2/2 \right| \le \dfrac{K_4}{4!}|x|^4$, where K_4 is a constant. Thus,

$$\left| \frac{\cos x - 1 + x^2/2}{x^3} \right| \le \frac{K_4 |x|^4}{4!\,|x|^3} = \frac{K_4}{4!}|x|.$$

Thus, $\lim\limits_{x \to 0} \dfrac{\cos x - 1 + x^2/2}{x^3} = 0.$

15. (a) $\sqrt{1 + x} \approx 1 + \dfrac{1}{2}x - \dfrac{1}{8}x^2.$

(b) $V = 2\pi\sigma\left(\sqrt{r^2 + a^2} - r \right) = 2\pi\sigma r \left(\sqrt{1 + \left(\dfrac{a}{r}\right)^2} - 1 \right)$

$$\approx 2\pi\sigma r \left(\frac{a^2}{2r^2} - \frac{a^4}{8r^4} \right) = 2\pi\sigma \left(\frac{a^2}{2r} - \frac{a^4}{8r^3} \right)$$

17. (a) $f(1.5) \approx f(1) + f'(1) \cdot (1.5 - 1) + \dfrac{f''(1)}{2}(1.5 - 1)^2$

$$= 1 + 2 \cdot (0.5) + \frac{1}{4}(0.5)^2 = \frac{33}{16} = 2.0625.$$

(b) Since $f'''(x) = -\dfrac{3x^2}{(1 + x^3)^2}$, $K_3 = 3/4$. Therefore, the approximation error does not exceed $\dfrac{K_3}{3!}(1.5 - 1)^3 = \dfrac{1}{64} = 0.015625.$

19. (a) By the fundamental theorem of calculus, $\displaystyle\int_a^x f'(t)\,dt = f(x) - f(a)$, and the result follows.

(b) Let $u = f'(t)$, $dv = dt$, $du = f''(t)\,dt$ and $v = t - x$. (Letting $v = t$ is also possible, but $v = t - x$ works better here.) Then

$$f(x) = f(a) + f'(t)(t - x)\Big|_a^x - \int_a^x (t - x)f''(t)\,dt$$

$$= f(a) - f'(a)(a - x) + \int_a^x (x - t)f''(t)\,dt.$$

(c) Let $u = f''(t)$ and $dv = (x - t)\,dt$. Then, $du = f'''(t)\,dt$ and $v = -(x - t)^2/2$. Thus,

$$f(x) = f(a) + f'(a)(x - a) - \frac{1}{2}f''(t)(x - t)^2\Big|_a^x + \frac{1}{2}\int_a^x (x - t)^2 f'''(t)\,dt$$

$$= f(a) + f'(a)(x - a) + \frac{1}{2}f''(a)(x - a)^2 + \frac{1}{2}\int_a^x (x - t)^2 f'''(t)\,dt.$$

This is the desired formula for $n = 2$.

Now, applying integration by parts with $u = f'''(t)$, $dv = (x - t)^2/2\,dt$, $du = f^{(4)}(t)\,dt$, $v = -(x - t)^3/3!$ gives (by a similar calculation)

$$f(x) = f(a) + f'(a)(x-a) + \frac{1}{2}f''(a)(x-a)^2 + \frac{1}{3!}f'''(a)(x-a)^3 + \frac{1}{3!}\int_a^x (x-t)^3 f^{(4)}(t)\,dt.$$

The same pattern persists for larger n.

§9.3 Fourier Polynomials: Approximating Periodic Functions

1. (a) Both $\sin(x)$ and $\cos(x)$ are 2π-periodic. Thus
    $f(x + 2\pi) = \cos(k(x + 2\pi)) = \cos(kx + 2k\pi) = \cos(kx) = f(x)$. The function g
    behaves the same way.

   (b) Each summand of a trigonometric polynomial is 2π-periodic. Therefore, so is the
    polynomial itself.

   (c) The smallest period of $p(x) = \cos(4x) + \sin(8x)$ is $\pi/2$, because
    $p(x + \pi/2) = \cos(4(x + \pi/2)) + \sin(8(x + \pi/2)) = \cos(4x + 2\pi) + \sin(8x + 4\pi) =$
    $\cos(4x) + \sin(8x) = p(x)$.

3. (a) Let $u = \sin(mx)$. Then, $\displaystyle\int_{-\pi}^{\pi} \cos(mx)\sin(mx)\,dx = \frac{1}{m}\int_{0}^{0} u\,du = 0$.

   (b) The integrand is odd because it's the product of an odd function and an even function.
    Integrating any odd function over $[-\pi, \pi]$ yields zero.

   (c) $\displaystyle\int_{-\pi}^{\pi} \cos(mx)\cos(nx)\,dx = \frac{1}{2}\int_{-\pi}^{\pi} \cos\big((m + n)x\big)\,dx$
    $$+ \frac{1}{2}\int_{-\pi}^{\pi} \cos\big((m - n)x\big)\,dx = 0 + 0$$
    by the result in part (a) of the previous exercise.

   (d) $\displaystyle\int_{-\pi}^{\pi} \cos^2(mx)\,dx = \left(\frac{x}{2} + \frac{\sin(2mx)}{4m}\right)\Bigg]_{-\pi}^{\pi} = \pi$.

   (e) $\displaystyle\int_{-\pi}^{\pi} \sin(mx)\sin(nx)\,dx = \frac{1}{2}\int_{-\pi}^{\pi} \cos\big((m - n)x\big)\,dx - \frac{1}{2}\int_{-\pi}^{\pi} \cos\big((m + n)x\big)\,dx$
    $$= 0 - 0$$
    by the result in part (a) of the previous exercise.

   (f) $\displaystyle\int_{-\pi}^{\pi} \sin^2(mx)\,dx = \left(\frac{x}{2} - \frac{\sin(2mx)}{4m}\right)\Bigg]_{-\pi}^{\pi} = \pi$.

5. All summands give zero integrals except the $a_k \cos(kx)$ term.

7. For all $k > 0$ we have $b_k = \dfrac{1}{\pi}\displaystyle\int_{-\pi}^{\pi} f(x)\,\sin(kx)\,dx$. Here the integrand is an odd function,
   since it's the product of an even function, $f(x)$, and an odd function, $\sin(kx)$. Integrating an
   odd function over $[-\pi, \pi]$ gives zero.

9. (a) The Fourier polynomial of degree 2 is $q_2(x) = 2/\pi$—the sum of terms through degree
    two in $q_5(x)$.

   (b) The Fourier polynomial of degree 3 is $q_3(x) = 2/\pi + 4\sin(3x)$—the sum of terms
    through degree three in $q_5(x)$.

   (c) $q_4(x) = 2/\pi + 4\sin(3x)$—the sum of terms through degree in $q_5(x)$.

11. The average value of f over $[-\pi, \pi]$ is $\dfrac{1}{2\pi} \displaystyle\int_{-\pi}^{\pi} f(x)\,dx$—exactly the same as the constant term a_0. Here the constant term is 3.

13. The key point is that each Fourier coefficient for the sum function $f + g$ is the sum of the corresponding Fourier coefficients for the separate functions f and g. To see why this works for the sine coefficients a_k, we'll write $a_{k,f}$, $a_{k,g}$, $a_{k,f+g}$ for the kth sine coefficient of f, g, and $f + g$, respectively. Then

$$a_{k,f+g} = \frac{1}{\pi} \int_{-\pi}^{\pi} (f(x) + g(x)) \sin(kx)\,dx$$

$$= \frac{1}{\pi} \int_{-\pi}^{\pi} f(x) \sin(kx)\,dx + \frac{1}{\pi} \int_{-\pi}^{\pi} g(x) \sin(kx)\,dx = a_{k,f} + a_{k,g}.$$

The situation is the same for the cosine coefficients: $b_{k,f+g} = b_{k,f} + b_{k,g}$.

15. (a) $a_0 = \dfrac{1}{2\pi} \displaystyle\int_{-\pi}^{\pi} f(x)\,dx = \dfrac{1}{2\pi} \displaystyle\int_{0}^{\pi} dx = \dfrac{1}{2}$

$a_k = \dfrac{1}{\pi} \displaystyle\int_{-\pi}^{\pi} f(x) \cos(kx)\,dx = \dfrac{1}{\pi} \displaystyle\int_{0}^{\pi} \cos(kx)\,dx = \dfrac{\sin(kx)}{k\pi} \Big]_{0}^{\pi} = 0$

$b_k = \dfrac{1}{\pi} \displaystyle\int_{-\pi}^{\pi} f(x) \sin(kx)\,dx = \dfrac{1}{\pi} \displaystyle\int_{0}^{\pi} \sin(kx)\,dx = -\dfrac{\cos(kx)}{k\pi} \Big]_{0}^{\pi} = \dfrac{1 - \cos(k\pi)}{k\pi}.$

Thus, $b_{2m} = 0$ and $b_{2m+1} = \dfrac{2}{(2m + 1)\pi}$ for $m = 0, 1, 2, 3, \ldots$.

(b) $q_1(x) = \dfrac{1}{2} + \dfrac{2 \sin x}{\pi}$;

$q_3(x) = \dfrac{1}{2} + \dfrac{2 \sin x}{\pi} + \dfrac{2 \sin(3x)}{3\pi}$;

$q_5(x) = \dfrac{1}{2} + \dfrac{2 \sin x}{\pi} + \dfrac{2 \sin(3x)}{3\pi} + \dfrac{2 \sin(5x)}{5\pi}$;

$q_7(x) = \dfrac{1}{2} + \dfrac{2 \sin x}{\pi} + \dfrac{2 \sin(3x)}{3\pi} + \dfrac{2 \sin(5x)}{5\pi} + \dfrac{2 \sin(7x)}{7\pi}$.

(d) Let s be the square wave function in Example 2. Then, $f(x) = \big(s(x) + 1\big)/2 = s(x)/2 + 1/2$. The same relationship holds between the Fourier polynomial found in Example 2 and the one in this exercise.

17. Because f is an even function, all the b_k are zero. To find the a_k, note that all integrands are even functions, so (for simplicity) we can integrate from 0 to π and double the result. Note, finally, that $f(x) = |x| = x$ if $x \geq 0$. Thus, we have $a_0 = \dfrac{1}{2\pi} \displaystyle\int_{-\pi}^{\pi} |x|\,dx = \dfrac{1}{\pi} \displaystyle\int_{0}^{\pi} x\,dx = \dfrac{\pi}{2}$. Similarly, if $k > 0$ we have $a_k = \dfrac{2}{\pi} \displaystyle\int_{0}^{\pi} x \cos(kx)\,dx$. Evaluating these integrals for $k = 1, 2, \ldots, 7$ gives the pattern

$$\frac{-4}{\pi}, \quad 0, \quad \frac{-4}{9\pi}, \quad 0, \quad \frac{-4}{25\pi}, \quad 0, \quad \frac{-4}{49\pi}.$$

Therefore, $q_7(x) = \dfrac{\pi}{2} - \dfrac{4}{\pi}\cos(x) - \dfrac{4}{9\pi}\cos(3x) - \dfrac{4}{25\pi}\cos(5x) - \dfrac{4}{49\pi}\cos(7x)$.

19. The constant function 4 is its own Fourier polynomial, and we've already found Fourier polynomials for x (which we can multiply by 3) and x^2 in earlier work. Combining these results gives

$$4 + \frac{\pi^2}{3} - 4\cos x + \cos(2x) - \frac{4}{9}\cos(3x) + \frac{1}{4}\cos(4x) - \frac{4}{25}\cos(5x) + \frac{1}{9}\cos(6x) - \frac{4}{49}\cos(7x)$$
$$+ 6\sin(x) - 3\sin(2x) + 2\sin(3x) - \frac{3}{2}\sin(4x) + \frac{6}{5}\sin(5x) - \sin(6x) + \frac{6}{7}\sin(7x).$$

10 Improper Integrals

§10.1 Improper Integrals: Ideas and Definitions

1. The interval of integration is infinite.

3. The integrand is unbounded near $x = 1$: $\lim\limits_{x \to 1^+} 1/(x^2 \ln x) = \infty$.

5. The integrand is improper at the left endpoint, so $\displaystyle\int_0^1 \frac{dx}{\sqrt{x}} = \lim\limits_{t \to 0^+} \int_t^1 \frac{dx}{\sqrt{x}}$ if the limit exists.

   But $\displaystyle\int_t^1 \frac{dx}{\sqrt{x}} = 2\sqrt{x}\Big]_t^1 = 2 - 2\sqrt{t}$. Thus, $\displaystyle\int_0^1 \frac{dx}{\sqrt{x}} = \lim\limits_{t \to 0^+} (2 - 2\sqrt{t}) = 2$.

7. $\displaystyle\int_1^\infty \frac{dx}{x^3} = \lim\limits_{t \to \infty} \int_1^t \frac{dx}{x^3} = \lim\limits_{t \to \infty} \frac{1}{2}\left(1 - \frac{1}{t^2}\right) = \frac{1}{2}$.

9. Integration by parts (with $u = \ln x$, $dv = dx/x^2$) gives $\displaystyle\int_1^t \frac{\ln x}{x^2}\, dx = 1 - \frac{1}{t} - \frac{\ln t}{t}$, so

   $\displaystyle\int_1^\infty \frac{\ln x}{x^2}\, dx = \lim\limits_{t \to \infty} \left(1 - \frac{1}{t} - \frac{\ln t}{t}\right) = 1$. (The last part of the limit can be found by l'Hôpital's rule.)

11. $\displaystyle\int_0^\infty e^{-x}\, dx = \lim\limits_{t \to \infty} \int_0^t e^{-x}\, dx = \lim\limits_{t \to \infty} -e^{-x}\Big]_0^t = \lim\limits_{t \to \infty} \left(1 - e^{-t}\right) = 1$.

13. $\displaystyle\int_{-\infty}^1 e^x\, dx = \lim\limits_{t \to -\infty} \left(e^1 - e^t\right) = e$.

15. The integrand is undefined at $x = 0$, so we want $\lim\limits_{t \to 0^+} \displaystyle\int_t^{16} \frac{dx}{\sqrt[4]{x^3}}$. Note that $\dfrac{1}{\sqrt[4]{x^3}} = \dfrac{1}{x^{3/4}}$;

   now a routine calculation gives $\displaystyle\int_t^{16} \frac{dx}{x^{3/4}} = 8 - 4\sqrt[4]{t}$; taking the limit as $t \to 0^+$ gives 8, the desired integral.

17. We want $\lim\limits_{t \to \infty} \displaystyle\int_3^t \frac{x}{\left(x^2 - 4\right)^3}\, dx$. The u-substitution $u = x^2 - 4$ leads to

   $\displaystyle\int_3^t \frac{x}{\left(x^2 - 4\right)^3}\, dx = -\frac{1}{4(x^2 - 4)^2}\Big]_3^t = \frac{1}{100} - \frac{1}{4(t^2 - 4)^2}$. Taking the limit as $t \to \infty$ gives $1/100$, the desired integral.

19. Yes, because $\displaystyle\int_0^\infty f(x)\, dx - \int_0^{100} f(x)\, dx = \int_{100}^\infty f(x)\, dx$, the right side is the difference of two convergent integrals, and so must converge itself.

21. The area between the x-axis and the curve $y = g(x)$ from $x = 1$ to $x = t$ is no more than the area between the x-axis and the curve $y = x^{-2}$ from $x = 1$ to $x = t$. The inequality implies that the first area is no greater than the second. The second area is found by integration: $\int_1^t x^{-2}\,dx = 1 - t^{-1}$, and this tends to 1 as $t \to \infty$. Hence the smaller area $\int_1^t g(x)\,dx$ must also tend to some finite limit (no bigger than 1) as $t \to \infty$.

23. Because f is odd we have $f(-x) = -f(x)$; it follows that $\int_{-\infty}^0 f(x)\,dx = -17$. Thus
$$\int_{-\infty}^\infty f(x)\,dx = \int_{-\infty}^0 f(x)\,dx + \int_0^\infty f(x)\,dx = 17 - 17 = 0.$$

25. (a) I is improper because the integrand is undefined at $x = 0$, (and blows up in magnitude nearby).

    (b) No; I diverges. For $\int_{-1}^1 \dfrac{1}{x^3}\,dx$ to converge requires that both improper integrals $\int_{-1}^0 \dfrac{1}{x^3}\,dx$ and $\int_0^1 \dfrac{1}{x^3}\,dx$ converge. In fact, both *diverge*, as easy calculations show.

27. (a) Substituting $u = \sqrt{x}$ and $du = \dfrac{1}{2\sqrt{x}}\,dx$ shows that
$$\int_1^t \frac{\cos(\sqrt{x})}{\sqrt{x}}\,dx = 2\sin(\sqrt{t}) - 2\sin 1.$$ Now $\sin(\sqrt{t})$ has no limit as $t \to \infty$ (see Example 6) so the original integral diverges.

    (b) The integral $\int_0^1 \dfrac{\cos(\sqrt{x})}{\sqrt{x}}\,dx$ converges. If $t > 0$, then (as above)
$$\int_t^1 \frac{\cos(\sqrt{x})}{\sqrt{x}}\,dx = 2\sin 1 - 2\sin(\sqrt{t}),$$ which tends to $2\sin 1$ as $t \to 0^+$.

29. The region S has infinite area — the improper integral $\int_0^1 \left(\dfrac{1}{x} - \dfrac{1}{\sqrt{x}} \right) dx$ diverges. To see why, note that (by straightforward calculations) $\int_t^1 \left(\dfrac{1}{x} - \dfrac{1}{\sqrt{x}} \right) dx = 2\sqrt{t} - \ln t - 2$, which tends to infinity as $t \to 0$.

31. The solid's cross-sectional area (perpendicular to the x-axis) at any given x is $\pi r^2 = \dfrac{\pi}{x^2}$. The volume is therefore given by the improper integral $\int_1^\infty \dfrac{\pi}{x^2}$; the value is π, as we calculated in Example 1.

33. Diverges. $\displaystyle\int_0^\infty \frac{x}{\sqrt{1+x^2}}\,dx = \lim_{t\to\infty}\left(\sqrt{1+t^2} - 1\right) = \infty.$

35. Converges.

$$\int_{-2}^{2} \frac{2x+1}{\sqrt[3]{x^2+x-6}}\, dx = \lim_{t \to 2^-} \int_{-2}^{t} \frac{2x+1}{\sqrt[3]{x^2+x-6}}\, dx = \lim_{t \to 2^-} \frac{3}{2}\left(x^2+x-6\right)^{2/3}\Big]_{-2}^{t}$$

$$= \lim_{t \to 2^-} \frac{3}{2}\left(\left(t^2+t-6\right)^{2/3} - (-4)^{2/3}\right) = -\frac{3\sqrt[3]{16}}{2} = -3\sqrt[3]{2}.$$

37. Converges.

$$\int_{2}^{4} \frac{x}{\sqrt{|x^2-9|}}\, dx = \int_{2}^{3} \frac{x}{\sqrt{9-x^2}}\, dx + \int_{3}^{4} \frac{x}{\sqrt{x^2-9}}\, dx$$

$$= \lim_{s \to 3^-} -\sqrt{9-x^2}\Big]_{2}^{s} + \lim_{t \to 3^+} \sqrt{x^2-9}\Big]_{t}^{4} = \sqrt{5} + \sqrt{7}.$$

39. Converges. $\displaystyle \int_{2}^{3} \frac{x}{\sqrt{3-x}}\, dx = \lim_{t \to 3^-} \left(\frac{2}{3}(3-t)^{3/2} - 6\sqrt{3-t} + \frac{16}{3}\right) = \frac{16}{3}.$

41. Diverges.

$$\int_{1}^{\infty} \frac{dx}{x(\ln x)^2} = \int_{1}^{2} \frac{dx}{x(\ln x)^2} + \int_{2}^{\infty} \frac{dx}{x(\ln x)^2}$$

$$= \lim_{s \to 1^+} \left(\frac{1}{\ln s} - \frac{1}{\ln 2}\right) + \lim_{t \to \infty} \left(\frac{1}{\ln 2} - \frac{1}{\ln t}\right) = \infty.$$

43. Diverges. $\displaystyle \int_{0}^{\infty} \frac{dx}{e^x-1} = \int_{0}^{1} \frac{dx}{e^x-1} + \int_{1}^{\infty} \frac{dx}{e^x-1} = \infty.$

45. Converges. $\displaystyle \int_{-\infty}^{\infty} \frac{dx}{e^x+e^{-x}} = \int_{-\infty}^{\infty} \frac{e^x}{e^{2x}+1}\, dx = \frac{\pi}{2}.$

47. Converges. $\displaystyle \int_{0}^{\pi/2} \frac{\cos x}{\sqrt{\sin x}}\, dx = 2.$

49. (a) If $p > 1$, $\displaystyle \int_{1}^{\infty} \frac{dx}{x^p} = \lim_{t \to \infty} \frac{1-t^{1-p}}{p-1} = \frac{1}{p-1}.$

   (b) If $p < 1$, $\displaystyle \int_{1}^{\infty} \frac{dx}{x^p} = \lim_{t \to \infty} \frac{1-t^{1-p}}{p-1} = \infty.$

51. Substituting $u = \ln x$ and $du = \dfrac{dx}{x}$ produces the equivalent integral $\displaystyle \int_{0}^{1} \frac{du}{u^p}$, which converges for $p < 1$ (by problem 50).

53. The integral diverges for all p. This is quite clear if $p \geq 0$, but it's true for negative p as well. One reason is that $x^p e^x \to \infty$ as $x \to \infty$ for all p.

55. (a) The integrand is unbounded near $x = 0$.

   (b) l'Hôpital's rule gives $\displaystyle \lim_{x \to 0^+} \frac{\sin x}{x} = 1$, so the integrand is actually bounded on the whole interval of integration.

57. Note that $f(x) = x/\ln x \to 0$ as $x \to 0^+$. Thus f is bounded, so the integral is proper.

59. $\displaystyle\int_0^\infty \left(\frac{2x}{x^2+1} - \frac{C}{2x+1} \right) dx = \ln(x^2+1) - \frac{C}{2}\ln(2x+1) = \ln\left(\frac{x^2+1}{\sqrt{(2x+1)^C}} \right)$. Thus, the improper integral converges to $-2\ln 2$ if $C = 4$ and diverges for all other values of C.

61. $\displaystyle\int_1^\infty \frac{x}{x^3+1}\,dx = \lim_{t\to\infty} \int_1^t \frac{x}{x^3+1}\,dx = \lim_{t\to\infty} -\int_1^{1/t} \frac{du}{1+u^3} = \int_0^1 \frac{du}{1+u^3}.$

63. Substituting $u = 1/x$, $du = -dx/x^2$, and $dx = -du/u^2$ (in the integrand and in the limits of integration) gives $\displaystyle\int_1^0 -\frac{du}{u^2(1+1/u^4)} = \int_0^1 \frac{u^2}{u^4+1}\,du.$

65. Using the substitution $u = 1/x$, $\displaystyle\int_1^\infty \frac{xe^{-1/x}}{1+x^4}\,dx = -\int_1^0 \frac{ue^{-u}}{1+u^4}\,du = \int_0^1 \frac{ue^{-u}}{1+u^4}\,du.$

   Alternatively, using the substitution $u = -1/x$, $\displaystyle\int_1^\infty \frac{xe^{-1/x}}{1+x^4}\,dx = -\int_{-1}^0 \frac{ue^{u}}{1+u^4}\,du.$

67. Let $u = 1/x$. Then $x = 1/u$ and $dx = -du/u^2$, and so

$$\int_0^1 \frac{dx}{\sqrt{x^4}\sqrt{x^{-1}+x^{-3}}} = \lim_{t\to 0^+} \int_t^1 \frac{dx}{x^2\sqrt{x^{-1}+x^{-3}}} \to \lim_{t\to 0^+} -\int_{1/t}^1 \frac{du}{u^2\sqrt{u^{-1}+u^{-3}}}$$

$$= \lim_{t\to 0^+} \int_1^{1/t} \frac{du}{\sqrt{u^3+u}}$$

$$= \int_1^\infty \frac{du}{\sqrt{u^3+u}} = \int_1^\infty \frac{dx}{\sqrt{x+x^3}}.$$

§10.2 Detecting Convergence, Estimating Limits

1. (a) $\dfrac{1}{g(x)} < \dfrac{1}{f(x)} < 1$.

    (b) $\dfrac{1}{(g(x))^r} < \dfrac{1}{(f(x))^r} < 1$; note that this holds for any $r \geq 1$.

    (c) $\dfrac{1}{(g(x))^r} < \dfrac{1}{(f(x))^r} < 1$, just as in (b).

3. (a) $f(x) = \dfrac{x^2 + 2}{x^2 + 1} = \dfrac{x^2 + 1 + 1}{x^2 + 1} = 1 + \dfrac{1}{x^2 + 1} > 1$.

    (b) The inequality in part (a) shows that $\displaystyle\int_{-\infty}^{\infty} f(x)\,dx > \int_{-\infty}^{\infty} 1\,dx$; the second integral is easily seen to diverge to infinity.

5. (a) For every real number x, we have $-1 \leq \sin x \leq 1$, which implies that $x - 1 \leq f(x) \leq x + 1$ for all x.

    (b) The inequality $x - 1 \leq f(x)$ implies that $\displaystyle\int_{2}^{\infty} (x - 1)\,dx \leq \int_{2}^{\infty} (x + \sin x)\,dx$. The left-hand integral is easily shown to diverge to infinity, and so must the right-hand integral.

7. (a) $0 \leq \sqrt{x}$ for all $x \geq 0$, so $x^2 \leq x^2 + \sqrt{x} = f(x)$ for all $x \geq 0$.

    (b) Part (a) implies that $\displaystyle\int_{1}^{\infty} \dfrac{dx}{f(x)} \leq \int_{1}^{\infty} \dfrac{dx}{x^2} = 1$; it follows that $\displaystyle\int_{1}^{\infty} \dfrac{dx}{f(x)}$ converges (to a value less than 1).

9. Diverges; $\displaystyle\int_{2}^{\infty} \dfrac{dx}{x - \sqrt{x}} > \int_{2}^{\infty} \dfrac{dx}{x}$, and the right-hand integral diverges.

11. $\displaystyle\int_{a}^{\infty} e^{-x}\,dx = e^{-a} \leq 10^{-5}$ if $a \geq \ln 100000 \approx 11.513$.

13. $\displaystyle\int_{a}^{\infty} \dfrac{dx}{x^2 + 1} = \dfrac{\pi}{2} - \arctan a \leq 10^{-5}$ if $a \geq \tan(\frac{\pi}{2} - 10^{-5}) \approx 100,000$.

15. $0 < \displaystyle\int_{a}^{\infty} \dfrac{dx}{x^2 + e^x} < \int_{a}^{\infty} e^{-x}\,dx = e^{-a} \leq 10^{-5}$ if $a \geq \ln(100000) \approx 11.513$. Thus, $\displaystyle\int_{0}^{12} \dfrac{dx}{x^2 + e^x}$ approximates $\displaystyle\int_{0}^{\infty} \dfrac{dx}{x^2 + e^x}$ within 10^{-5}.

17. $0 < \displaystyle\int_{a}^{\infty} \dfrac{\arctan x}{(1 + x^2)^3}\,dx < \dfrac{\pi}{2} \int_{a}^{\infty} \dfrac{dx}{x^6} = \dfrac{\pi}{10a^5} \leq 10^{-5}$ if $a \geq (10^4 \pi)^{1/5} \approx 7.9329$. Thus, $\displaystyle\int_{0}^{8} \dfrac{\arctan x}{(1 + x^2)^3}\,dx$ approximates $\displaystyle\int_{0}^{\infty} \dfrac{\arctan x}{(1 + x^2)^3}\,dx$ within 10^{-5}.

19. Let $I = \int_1^\infty f(x)\,dx$. Then, if $a > 1$, $I - \int_1^a f(x)\,dx = \int_a^\infty f(x)\,dx$. Since
    $\lim\limits_{a\to\infty} \int_1^a f(x)\,dx = I$, $\lim\limits_{a\to\infty} \int_a^\infty f(x)\,dx = 0$. This implies that if a is large enough,
    $\int_a^\infty f(x)\,dx \leq 10^{-10}$.

21. $I = \int_0^\infty f(x)\,dx = \int_0^a f(x)\,dx + \int_a^\infty f(x)\,dx \implies \left| I - \int_0^a f(x)\,dx \right|$
    $= \left| \int_a^\infty f(x)\,dx \right| \leq 0.0001$.

23. $\int_0^\infty e^{-x^2}\,dx = \int_0^1 e^{-x^2}\,dx + \int_1^\infty e^{-x^2}\,dx$. Since the first integral is proper and the second
    is known to converge, the original improper integral must also converge.

25. The hint implies that $(\ln x)^p < x$ for large x. This implies that the given integral is
    comparable to $\int_2^\infty \frac{dx}{x}$, which diverges. Therefore I diverges for *all* positive constants p.

27. Diverges: Since $x^4 \leq x$ if $0 \leq x \leq 1$,
    $\int_0^\infty \frac{dx}{x^4 + x}\,dx = \int_0^1 \frac{dx}{x^4 + x}\,dx + \int_1^\infty \frac{dx}{x^4 + x}\,dx \geq \int_0^1 \frac{dx}{2x}\,dx + \int_1^\infty \frac{dx}{x^4 + x}\,dx$. Since
    $\int_0^1 \frac{dx}{2x}\,dx$ diverges, the original improper integral also diverges.

29. Converges: $0 \leq \int_1^\infty \frac{dx}{\sqrt[3]{x^6 + x}} \leq \int_1^\infty \frac{dx}{x^2} = 1$.

31. $0 < \int_0^\infty \frac{e^{-x}}{\sqrt{x}}\,dx = \int_0^1 \frac{e^{-x}}{\sqrt{x}}\,dx + \int_1^\infty \frac{e^{-x}}{\sqrt{x}}\,dx \leq \int_0^1 \frac{dx}{\sqrt{x}} + \int_1^\infty e^{-x}\,dx = 2 + e^{-1}$. Both of
    the last two summands are convergent integrals, so the original improper integral converges.

33. (a) Since $\left| \frac{\cos x}{x^2} \right| \leq \frac{1}{x^2}$ and $\int_1^\infty x^{-2}\,dx$ converges, Theorem 2 says that $\int_1^\infty \frac{\cos x}{x^2}\,dx$
    converges.

    (b) Let $u = x^{-1}$ and $dv = \sin x\,dx$. Then,
    $\int_1^\infty \frac{\sin x}{x}\,dx = -\frac{\cos x}{x} \Big]_1^\infty - \int_1^\infty \frac{\cos x}{x^2}\,dx = \cos 1 - \int_1^\infty \frac{\cos x}{x^2}\,dx$. Thus,
    $\int_1^\infty \frac{\sin x}{x}\,dx$ can be written as the difference of two finite numbers.

35. By l'Hôpital's rule, $\lim\limits_{x\to\infty} \frac{\int_1^x \sqrt{1 + e^{-3t}}\,dt}{x} = \lim\limits_{x\to\infty} \frac{\sqrt{1 + e^{-3x}}}{1} = 1$. [NOTE: The improper
    integral in the numerator diverges to ∞ because $\sqrt{1 + e^{-3t}} > 1$ for all $t \geq 1$. This assures
    that l'Hôpital's rule does apply.]

§10.3   Improper Integrals and Probability

3. When $Z = \dfrac{x - 500}{100}, dZ = \dfrac{dx}{100}$. Furthermore, if $x = 500, Z = 0$; if $x = 700, Z = 2$. Thus,

$$I_1 = \frac{1}{100\sqrt{2\pi}} \int_{500}^{700} \exp\left(-\frac{(x-500)^2}{2 \cdot 100^2}\right) dx$$

$$= \frac{1}{100\sqrt{2\pi}} \int_0^2 \exp\left(-\frac{Z^2}{2}\right) 100\,dZ = \frac{1}{\sqrt{2\pi}} \int_0^2 e^{-Z^2/2}\,dZ.$$

5. (a) Since the integrand is an even function

$$\int_{-\infty}^{\infty} e^{-x^2}\,dx = 2\int_0^{\infty} e^{-x^2}\,dx = 2 \cdot \frac{\sqrt{\pi}}{2} = \sqrt{\pi}.$$

   (b) Using the substitution $u = x/\sqrt{2}$, $\dfrac{1}{\sqrt{2\pi}} \int_{-\infty}^{\infty} e^{\frac{-x^2}{2}} = \dfrac{1}{\sqrt{\pi}} \int_{-\infty}^{\infty} e^{-u^2}\,du = 1.$

   (c) Using the substitution $u = (x - m)/s$,

$$\frac{1}{\sqrt{2\pi}\,s} \int_{-\infty}^{\infty} e^{\frac{-(x-m)^2}{2s^2}} = \frac{1}{2\sqrt{\pi}} \int_{-\infty}^{\infty} e^{-u^2/2}\,du = 1.$$

7. (a) $z = \dfrac{600 - 500}{100} = 1.$

   (b) $z = \dfrac{450 - 500}{100} = -0.5.$

   (c) $\dfrac{1}{\sqrt{2\pi} \cdot 100} \displaystyle\int_{450}^{\infty} \exp\left(-\dfrac{(x-500)^2}{2 \cdot 100^2}\right) dx = \dfrac{1}{\sqrt{2\pi}} \displaystyle\int_{-0.5}^{\infty} e^{-x^2/2}\,dx.$

   (d) $\dfrac{1}{\sqrt{2\pi} \cdot 100} \displaystyle\int_{-\infty}^{600} \exp\left(-\dfrac{(x-500)^2}{2 \cdot 100^2}\right) dx = \dfrac{1}{\sqrt{2\pi}} \displaystyle\int_{-\infty}^{1} e^{-x^2/2}\,dx.$

   (e) $\dfrac{1}{\sqrt{2\pi} \cdot 100} \displaystyle\int_{450}^{600} \exp\left(-\dfrac{(x-500)^2}{2 \cdot 100^2}\right) dx = \dfrac{1}{\sqrt{2\pi}} \displaystyle\int_{-0.5}^{1} e^{-x^2/2}\,dx.$

9. (a) Since $n(-t) = n(t)$,

$$A(-z) = \int_{-\infty}^{-z} n(t)\,dt = \int_{z}^{\infty} n(t)\,dt = \int_{-\infty}^{\infty} n(t)\,dt - \int_{-\infty}^{z} n(t)\,dt = 1 - A(z).$$

   (b) $A(-1.2) = 1 - A(1.2) \approx 1 - 0.8849 = 0.1151.$

11. The probability that $-1.5 \le Z \le 0$ is the same as the probability that $0 \le Z \le 1.5$, which is $A(1.5) - A(0) \approx 0.9333 - 0.5 = 0.4333$. Thus about 43% of scores fall in this range.

13. $z = (4 - 10)/5 = -1.2$ and $A(-1.2) = 1 - A(1.2) \approx 0.1151$. Thus,

$$\frac{1}{5\sqrt{2\pi}} \int_{-\infty}^{4} \exp\left(-\frac{(x-10)^2}{50}\right) \approx 0.1151.$$

15. Counting rectangles under the graph between $x = 0.5$ and $x = 1$ gives a bit more than 8 rectangles, or just over 0.4 in area.

17. Counting rectangles under the graph to the right of $x = 1.5$ gives about 3 rectangles, or about 0.15 in area.

19. The Z-scores that corresponds to differing from the average by more than 1 inch are $Z = \pm 1/0.6 \approx \pm 1.67$. Thus, the probability is (approximately)

$$\int_{-\infty}^{-1.67} n(x)\,dx + \int_{1.67}^{\infty} n(x)\,dx \approx 0.095581.$$

21. Let D be the "nominal" diameter at which the machine is set. Then the Z-score of a can top with diameter greater than 3 inches is greater than $Z_0 = (3 - D)/0.01$. No more than 10% of the can tops produced will have diameter greater than 3 inches if $\int_{Z_0}^{\infty} n(x)\,dx = 0.1$, that is if $Z_0 \approx 1.3$ (from the table). This corresponds to $D = 2.987$ inches.

23. (a) If $\beta < 0$, then g is not a positive function. If $\beta = 0$, then $g(x)$ is undefined. Therefore, $\beta > 0$ must be true for g to be a probability distribution.

    (b) If $u = (\ln x - \alpha)/\beta$, then $du = \big(1/(\beta x)\big)\,dx$, so

    $$\int_a^b g(x)\,dx \rightarrow \int_{(\ln a - \alpha)/\beta}^{(\ln b - \alpha)/\beta} \frac{1}{\sqrt{2\pi}} e^{-u^2/2}\,du$$

    $$= \int_{-\infty}^{(\ln b - \alpha)/\beta} \frac{1}{\sqrt{2\pi}} e^{-u^2/2}\,du - \int_{-\infty}^{(\ln a - \alpha)/\beta} \frac{1}{\sqrt{2\pi}} e^{-u^2/2}\,du$$

    $$= A\left(\frac{\ln b - \alpha}{\beta}\right) - A\left(\frac{\ln a - \alpha}{\beta}\right).$$

25. Since $\int_{-\infty}^{\infty} f(x)\,dx = k\pi$, f is a probability density function if $k = 1/\pi$.

27. (a) f is a probability density function because it is a positive function and

    $$\int_{-\infty}^{\infty} f(x)\,dx = \int_0^{\infty} \lambda e^{-\lambda x}\,dx = -e^{-\lambda x}\Big]_0^{\infty} = 1.$$

    (b) $\int_{-\infty}^{\infty} x f(x)\,dx = \lambda \int_0^{\infty} x e^{-\lambda x}\,dx = -\lambda \frac{1 + \lambda x}{\lambda^2} e^{-\lambda x}\Big]_0^{\infty} = \frac{1}{\lambda}.$

29. (a) $\Gamma(1) = \int_0^{\infty} e^{-t}\,dt = \lim_{a \to \infty} \int_0^a e^{-t}\,dt = \lim_{a \to \infty} \left(1 - e^{-a}\right) = 1.$

    (b) If $x > 1$, an integration by parts (with $u = t^{x-1}$ and $dv = e^{-t}\,dt$) shows that

    $$\Gamma(x) = \int_0^{\infty} t^{x-1} e^{-t}\,dt = -t^{x-1} e^{-t}\Big]_0^{\infty} + (x - 1)\int_0^{\infty} t^{x-2} e^{-t}\,dt = 0 + (x - 1)\Gamma(x - 1).$$

31. The substitution $u = \sqrt{t}$ shows that $\Gamma\left(\frac{1}{2}\right) = \int_0^{\infty} \frac{dt}{\sqrt{t}\,e^t} = 2\int_0^{\infty} e^{-u^2}\,du = \sqrt{\pi}.$

33. First make the substitution $x = e^{-u}$, then make the substitution $w = (m + 1)u$:

$$\int_0^1 x^m (\ln x)^n \, dx \rightarrow -\int_\infty^0 e^{-(m+1)u}(-u)^n \, du = (-1)^n \int_0^\infty u^n e^{-(m+1)u} \, du$$

$$\rightarrow \frac{(-1)^n}{(m+1)^{n+1}} \int_0^\infty w^n e^{-w} \, dw = \frac{(-1)^n \, n!}{(m+1)^{n+1}}.$$

35. The entire sample space is the square $[-2, 2] \times [-2, 2]$, which has area $4^2 = 16$. The probability that $x^2 + y^2 \leq 1$ is the proportion of the total area occupied by the interior of the circle $x^2 + y^2 \leq 1$. This circle has area π, so the probability is $\pi/16 \approx 0.196$.

37. The sample space is the square $[0, 1] \times [0, 1]$, which has area 1. The probability that $y/x \geq 2$ is the proportion of the total area occupied by the triangle with corners $(0, 0)$, $(0, 1)$, and $(1/2, 1)$. (In this triangle, $y/x \geq 2$.) This triangle has area $1/4$; this is the desired proportion.)

39. The sample space is the square $[0, 3] \times [0, 3]$, which has area 9. The probability that $x + y \geq xy$ is the proportion of the total area occupied by the part of the square that lies below the curve $y = x/(x - 1)$. A look at graphs shows that this part of the square has area

$$\frac{9}{2} + \int_{3/2}^3 \frac{x}{x - 1} \, dx = 6 + \ln 4 \approx 7.386,$$ so the desired probability is $(6 + \ln 4)/9 \approx 0.821$.

11   *Infinite Series*

§11.1   Sequences and Their Limits

1. $\lim\limits_{k\to\infty} a_k$ does not exist: $\lim\limits_{k\to\infty} a_{2k} = \infty$ but $\lim\limits_{k\to\infty} a_{2k+1} = -\infty$.

3. $\lim\limits_{k\to\infty} a_k = \infty$.

5. $\lim\limits_{k\to\infty} a_k = 0$.

7. $\lim\limits_{k\to\infty} a_k = \pi/2$ because $\lim\limits_{x\to\infty} \arctan x = \pi/2$.

9. $\lim\limits_{k\to\infty} a_k = 0$.

11. $\lim\limits_{k\to\infty} a_k = 1$.

13. $\lim\limits_{m\to\infty} a_m = \lim\limits_{m\to\infty} \dfrac{m^2}{e^m} = \lim\limits_{m\to\infty} \dfrac{2m}{e^m} = \lim\limits_{m\to\infty} \dfrac{2}{e^m} = 0$.

15. Since $\dfrac{k!}{(k+1)!} = \dfrac{1}{k+1}$, $\lim\limits_{k\to\infty} a_k = 0$.

17. $\lim\limits_{k\to\infty} \ln a_k = \lim\limits_{k\to\infty} \dfrac{\ln 3}{k} = 0$. Therefore, $\lim\limits_{k\to\infty} a_k = 1$.

19. $a_k = (-1)^k/k$. $\lim\limits_{k\to\infty} a_k = 0$, but $\{a_k\}$ is neither an increasing nor a decreasing sequence.

21. $a_k = k$; $a_k < a_{k+1}$ for all integers $k \geq 1$, but $\lim\limits_{k\to\infty} a_k = \infty$.

23. $a_k = e^{-k}$; $a_{k+1} < a_k$ for all integers $k \geq 1$, and $\lim\limits_{k\to\infty} a_k = 0$.

25. The sequence is bounded above (e.g, by 0.8) and is monotone increasing and by Theorem 3 has a limit.

27. $\lim\limits_{n\to\infty} a_n = \lim\limits_{n\to\infty} n\sin(1/n) = \lim\limits_{n\to\infty} \dfrac{\sin\left(n^{-1}\right)}{n^{-1}} = \lim\limits_{n\to\infty} \cos\left(n^{-1}\right) = 1$.

29. $\lim\limits_{k\to\infty} a_k = \displaystyle\int_0^\infty e^{-x}\,dx = 1$.

31. $a_k = \sqrt{k^2+1} - k = \dfrac{\left(\sqrt{k^2+1}-k\right)\cdot\left(\sqrt{k^2+1}+k\right)}{\sqrt{k^2+1}+k} = \dfrac{1}{\sqrt{k^2+1}+k}$. Thus, $\lim\limits_{k\to\infty} a_k = 0$.

33. $\lim\limits_{k\to\infty} a_k$ exists only when $x \leq 0$ because $e^x > 1$ when $x > 0$. When $x < 0$, $\lim\limits_{k\to\infty} a_k = 0$. When $x = 0$, $\lim\limits_{k\to\infty} a_k = 1$.

35. $\lim\limits_{k\to\infty} a_k = 0$ when $-\sin 1 < x < \sin 1$ and $\lim\limits_{k\to\infty} a_k = 1$ when $x = \sin 1$.

37. $\lim\limits_{n\to\infty} \dfrac{\ln x}{n} = 0 \implies \lim\limits_{n\to\infty} x^{1/n} = e^0 = 1$ for all $x > 0$.

39. Let $a_n = \left(1 - \dfrac{1}{2n}\right)^n$. Then $\ln(a_n) = n \ln\left(1 - \dfrac{1}{2n}\right) = \dfrac{\ln\left(1 - \frac{1}{2n}\right)}{\frac{1}{n}}$ so, by l'Hôpital's rule,

$\lim_{n\to\infty} \ln(a_n) = -1/2$. It follows that $\lim\limits_{n\to\infty} a_n = e^{-1/2}$.

41. (a) $a_{n+1} = \displaystyle\sum_{k=1}^{n+1} \dfrac{1}{(n+1)+k} = \sum_{k=2}^{n+2} \dfrac{1}{n+k} = \sum_{k=2}^{n} \dfrac{1}{n+k} + \dfrac{1}{2n+1} + \dfrac{1}{2n+2}$

$> \displaystyle\sum_{k=2}^{n} \dfrac{1}{n+k} + \dfrac{2}{2n+2} = \sum_{k=1}^{n} \dfrac{1}{n+k} = a_n.$

(b) $a_n = \displaystyle\sum_{k=1}^{n} \dfrac{1}{n+k} \le \sum_{k=1}^{n} \dfrac{1}{n+1} = \dfrac{n}{n+1} < 1$ when $n \ge 1$.

(c) Parts (a) and (b) imply that the sequence $\{a_n\}$ is monotonically increasing and bounded above. Therefore, $\lim\limits_{n\to\infty} a_n$ exists.

(d) Part (a) implies that the sequence $\{a_n\}$ is monotonically increasing. Therefore, $\lim\limits_{n\to\infty} a_n > a_k$ for any integer $k \ge 1$. Since $a_1 = 1/2$, $\lim\limits_{n\to\infty} a_n > 1/2$.

(e) $\lim\limits_{n\to\infty} a_n = \displaystyle\int_0^1 \dfrac{dx}{1+x} = \ln 2.$

43. (a) First, note that $\ln(\sqrt[n]{n!}) = \ln(n!)^{1/n} = \dfrac{1}{n}\ln(n!) = \dfrac{1}{n}\displaystyle\sum_{k=1}^{n} \ln k$. Also, $n = (n^n)^{1/n}$ so

$\ln n = \dfrac{1}{n}\ln(n^n) = \dfrac{1}{n}\displaystyle\sum_{k=1}^{n} \ln n$. Therefore,

$\ln a_n = \ln\left(\dfrac{\sqrt[n]{n!}}{n}\right) = \ln\left(\sqrt[n]{n!}\right) - \ln n = \dfrac{1}{n}\displaystyle\sum_{k=1}^{n} \ln k - \dfrac{1}{n}\sum_{k=1}^{n} \ln n.$

(b) The right sum approximation to $\displaystyle\int_0^1 \ln x \, dx$ is

$$R_n = \dfrac{1}{n}\sum_{k=1}^{n} \ln\left(\dfrac{k}{n}\right) = \dfrac{1}{n}\sum_{k=1}^{n} \ln k - \dfrac{1}{n}\sum_{k=1}^{n} \ln n = \ln a_n.$$

(c) The result in part (b) implies that $\lim\limits_{n\to\infty} \ln a_n = \displaystyle\int_0^1 \ln x \, dx = x \ln x - x \Big]_0^1 = -1.$

Therefore, $\lim\limits_{n\to\infty} a_n = e^{-1}.$

45. Observe that $\dfrac{n+3}{2n+1} \leq \dfrac{6}{7}$ for all $n \geq 3$. Now,

$$\frac{a_{n+1}}{a_3} = \frac{a_{n+1}}{a_n} \cdot \frac{a_n}{a_{n-1}} \cdots \frac{a_4}{a_3} \leq \left(\frac{6}{7}\right)^{n-2} \implies a_{n+1} \leq a_3(6/7)^{n-2} \implies \lim_{n\to\infty} a_n = 0.$$

47. No. The terms of the sequence change sign.

49. Yes. $|a_{n+1}| < |a_n|$

51. Let $a_n = \sin\left(\pi/n^2\right)$. Then, the inequality $0 < \sin x < x$ when $0 < x < 1$ implies that $0 < a_n < \pi/n^2$. Since $\lim_{n\to\infty} \pi/n^2 = 0$, Theorem 2 implies the result.

53. Yes. The inequalities $0 < a_{n+1} = a_n/2 < a_n$ imply that the sequence is monotonically decreasing and bounded below.

55. $a_{n+1} = x^{1/2^n}$ so $\lim_{n\to\infty} a_n = 1$ for all $x > 0$ by the previous problem. When $x = 0$, $\lim_{n\to\infty} a_n = 0$. Thus, $\lim_{n\to\infty} a_n$ exists for all $x \geq 0$.

§11.2 Infinite Series, Convergence, and Divergence

1. (a) $a_1 = 1/5, a_2 = 1/25, a_5 = 1/3125, a_{10} = 1/9765625.$

$$S_n = \sum_{k=0}^{n} \frac{1}{5^k} = \frac{1 - (1/5)^{n+1}}{1 - (1/5)} = \frac{5 - (1/5)^n}{4}. \text{ Thus, } S_1 = 6/5 = 1.2,$$

$S_2 = 31/25 = 1.24, S_5 = 3906/3125 = 1.24992,$ and
$S_{10} = 12207031/9765625 = 1.2499999744.$

(b) For any $k \geq 0, 0 < a_{k+1} = \dfrac{1}{5^{k+1}} = \dfrac{1}{5}\dfrac{1}{5^k} = \dfrac{a_k}{5} < a_k.$ Thus, the sequence $\{a_k\}$ is decreasing and bounded below.

(c) $S_{n+1} = S_n + 1/5^{n+1} > S_n,$ so the sequence is increasing. Since $S_n = \dfrac{5}{4} - \dfrac{1}{4}\left(\dfrac{1}{5}\right)^n,$ $S_n < 5/4$ for all $n \geq 0.$ Because the sequence of partial sums is increasing and bounded above, it must converge.

(d) $S_n = \dfrac{5}{4} - \dfrac{1}{4}\left(\dfrac{1}{5}\right)^n \implies \lim_{n\to\infty} S_n = \dfrac{5}{4}.$

(e) $R_1 = 1/20, R_2 = 1/100, R_5 = 1/12500, R_{10} = 1/39062500.$

(f) $R_n = \displaystyle\sum_{k=n+1}^{\infty} 5^{-k} = \dfrac{1}{4}\left(\dfrac{1}{5}\right)^n.$ Thus, $0 < R_{n+1} < R_n$ for all $n \geq 0.$

(g) $\lim_{n\to\infty} R_n = 0.$

3. (a) $a_{k+1} = \dfrac{1}{(k+1) + 2^{k+1}} < \dfrac{1}{k + 2^k} = a_k$ since $k < k + 1$ and $2^k < 2^{k+1}.$

(b) If $k \geq 0, a_k = \dfrac{1}{k + 2^k} \leq \dfrac{1}{2^k} = 2^{-k}.$

(c) $S_{n+1} = S_n + \dfrac{1}{(n+1) + 2^{n+1}} > S_n$ for all $n \geq 0.$

(d) $S_n = \displaystyle\sum_{k=0}^{n} a_k \leq \sum_{k=0}^{n} 2^{-k} = \dfrac{1 - \left(\frac{1}{2}\right)^{n+1}}{1 - \frac{1}{2}} = 2 - 2^{-n} < 2$ for all $n \geq 0.$

(e) From parts (c) and (d) we know that the sequence of partial sums is increasing and bounded above. Therefore, the sequence of partial sums converges (i.e., the series converges).

5. (a) If $k \geq 1, k! = 1 \cdot \underbrace{2 \cdot 3 \cdots (k-1) \cdot k}_{k-1 \text{ terms}} \geq 1 \cdot \underbrace{2 \cdot 2 \cdots 2 \cdot 2}_{k-1 \text{ terms}} = 2^{k-1}$ and so $\dfrac{1}{k!} \leq \dfrac{1}{2^{k-1}}.$

(b) $S_{n+1} = \displaystyle\sum_{k=0}^{n+1} a_k = \sum_{k=0}^{n} a_k + a_{n+1} = S_n + a_{n+1}.$ Since $a_{n+1} > 0, S_{n+1} > S_n.$

(c) Part (a) implies that

$$S_n = \sum_{k=0}^{n} \frac{1}{k!} \le 1 + \sum_{k=1}^{n} \frac{1}{2^{k-1}} = 1 + \sum_{k=0}^{n-1} \frac{1}{2^k} < 1 + \sum_{k=0}^{\infty} \frac{1}{2^k} = 1 + 2 = 3.$$ Since the

sequence of partial sums $\{S_n\}$ is increasing and bounded, the series converges.

7. $\displaystyle \sum_{i=0}^{\infty} \frac{1}{(i+1)^4} = \sum_{j=1}^{\infty} \frac{1}{j^4} = \frac{\pi^4}{90}.$

9. (a) When $r = 1$, $S_n = (n+1)a$, but the right-hand side of the expression is undefined (because of the zero in the denominator).

    (b) $S_n - rS_n = (a + ar + ar^2 + \cdots + ar^n) - r(a + ar + ar^2 + \cdots + ar^n) = a - ar^{n+1} = a(1 - r^{n+1}).$

    (c) $S_n - rS_n = (1-r)S_n = a(1 - r^{n+1})$. The desired result is now obtained by dividing through by $1 - r$.

11. $\displaystyle \frac{1}{16} + \frac{1}{32} + \frac{1}{64} + \frac{1}{128} + \cdots + \frac{1}{2^{i+4}} + \cdots = \frac{1}{2^4} \sum_{i=0}^{\infty} \left(\frac{1}{2}\right)^i = \frac{1}{2^4} \cdot 2 = \frac{1}{8}.$

13. $\displaystyle \sum_{n=0}^{\infty} e^{-n} = \frac{e}{e-1}.$

15. $\displaystyle \sum_{m=2}^{\infty} (\arctan 1)^m = \sum_{m=2}^{\infty} (\pi/4)^m = \left(\frac{\pi}{4}\right)^2 \sum_{m=0}^{\infty} (\pi/4)^m = \frac{\pi^2}{16} \frac{1}{1 - \pi/4} = \frac{\pi^2}{16 - 4\pi}.$

17. $\displaystyle \sum_{j=5}^{\infty} \left(-\frac{1}{2}\right)^j = \sum_{j=0}^{\infty} \left(-\frac{1}{2}\right)^j - \sum_{j=0}^{4} \left(-\frac{1}{2}\right)^j$

$$= \frac{1}{1 - (-1/2)} - \frac{1 - (-1/2)^5}{1 - (-1/2)} = -\frac{1}{48} \approx -0.020833.$$

19. Since $1/(2 + \sin k) \ge 1/3$ for all $k \ge 1$, the series diverges by the nth-term test.

21. $S_n = \arctan(n+1) - \arctan(0) = \arctan(n+1)$. Since $\displaystyle \lim_{n \to \infty} S_n = \frac{\pi}{2}$, the series converges to $\dfrac{\pi}{2}$.

23. $S_n = 1 + \dfrac{1}{\sqrt{2}} - \dfrac{1}{\sqrt{n+1}} - \dfrac{1}{\sqrt{n+2}}$ when $n \ge 1$. Since $\displaystyle \lim_{n \to \infty} S_n = 1 + 1/\sqrt{2}$, the series converges to $1 + 1/\sqrt{2}$.

25. (a) $S_{100} = 5 - \dfrac{3}{100} = 4.97.$

(b) $\displaystyle\sum_{k=1}^{\infty} a_k = \lim_{N\to\infty} \sum_{k=1}^{N} a_k = \lim_{N\to\infty} S_N = \lim_{N\to\infty}\left(5 - \frac{3}{N}\right) = 5.$

(c) $\displaystyle\lim_{k\to\infty} a_k = 0$ since $\displaystyle\sum_{k=1}^{\infty} a_k$ converges (Theorem 6).

(d) Since $a_1 = 2$, and $a_{n+1} = S_{n+1} - S_n = \left(5 - \dfrac{3}{n+1}\right) - \left(5 - \dfrac{3}{n}\right) = \dfrac{3}{n^2+n} > 0$ for

all $n \geq 1$, $a_k > 0$ for all $k \geq 1$.

27. $\dfrac{1}{4} + \dfrac{1}{16} + \dfrac{1}{36} + \dfrac{1}{100} + \cdots = \dfrac{1}{4}\left(1 + \dfrac{1}{4} + \dfrac{1}{9} + \cdots\right) = \dfrac{\pi^2}{24}.$

29. $\displaystyle\sum_{m=1}^{\infty} \frac{(-1)^{m+1}}{m^2} = \sum_{m=1}^{\infty} \frac{1}{m^2} - \frac{1}{2}\sum_{m=1}^{\infty}\frac{1}{m^2} = \frac{\pi^2}{12}.$

Alternatively, $\displaystyle\sum_{m=1}^{\infty}\frac{(-1)^{m+1}}{m^2} = \sum_{k=0}^{\infty}\frac{1}{(2k+1)^2} - \sum_{j=1}^{\infty}\frac{1}{(2j)^2} = \frac{\pi^2}{8} - \frac{\pi^2}{24} = \frac{\pi^2}{12}.$

31. The series converges for all values of x such that $-1 < x < 1$. $\displaystyle\sum_{k=0}^{\infty} x^k = \frac{1}{1-x}.$

33. Since $\displaystyle\sum_{j=5}^{\infty} x^{2j} = x^{10}\sum_{j=0}^{\infty}\left(x^2\right)^j$, the series converges when $-1 < x^2 < 1$. Thus, the series

converges to $\dfrac{x^{10}}{1-x^2}$ for all values of x such that $-1 < x < 1$.

35. The series converges when $|1+x| < 1$. Thus, the series converges for all values of x such

that $-2 < x < 0$. $\displaystyle\sum_{n=3}^{\infty}(1+x)^n = (1+x)^3\sum_{n=0}^{\infty}(1+x)^n = -\frac{(1+x)^3}{x}.$

37. $\displaystyle S_{n+1} = \sum_{k=0}^{n}(1/3)^k \implies \lim_{n\to\infty} S_n = 3/2.$

39. Diverges by the nth term test: $\displaystyle\lim_{n\to\infty}\frac{n+1}{2n+1} = \frac{1}{2} \neq 0.$

41. The partial sums of the series are

$\displaystyle S_N = \sum_{n=2}^{N}\frac{2}{n^2-1} = \sum_{n=2}^{N}\left(\frac{1}{n-1} - \frac{1}{n+1}\right) = 1 + \frac{1}{2} - \frac{1}{N} - \frac{1}{N+1}.$ Therefore, the series

converges to $3/2$.

43. Diverges by the nth term test: $\displaystyle\lim_{n\to\infty}\sqrt[n]{\pi} = 1 \neq 0.$

45. Converges—by the comparison test:

$$\sum_{j=2}^{\infty} \frac{3^j}{4^{j+1}} = \frac{1}{4} \sum_{j=2}^{\infty} \left(\frac{3}{4}\right)^j = \frac{9}{64} \sum_{j=0}^{\infty} \left(\frac{3}{4}\right)^j = \frac{9}{64} \cdot \frac{1}{1-3/4} = \frac{9}{16}.$$

$$R_N = \sum_{j=N+1}^{\infty} \frac{3^j}{4^{j+1}} = \frac{3^{N+1}}{4^{N+2}} \sum_{j=0}^{\infty} \left(\frac{3}{4}\right)^j = \left(\frac{3}{4}\right)^{N+1} \le 0.001 \text{ when}$$

$$N \ge \frac{\ln 0.001}{\ln 3/4} - 1 \approx 23.012. \text{ Thus, } n \ge 24 \text{ implies that } S_n \text{ approximates the sum of the}$$
series within 0.001.

47. Diverges—each term of this series is a constant multiple (1/100) of the corresponding term of the harmonic series.

49. The series $\sum_{k=0}^{\infty} \frac{3}{10} \left(-\frac{1}{2}\right)^k$ converges to 1/5.

51. Diverges by the nth term test.

53. After the first bounce, the ball rebounds to a height of $4 \cdot (2/3)$ feet. The ball then falls from this height and rebounds to $4 \cdot (2/3)^2$ feet, etc. Thus, the total distance that the ball travels is

$$4 + 2 \cdot 4 \cdot \frac{2}{3} + 2 \cdot 4 \cdot \left(\frac{2}{3}\right)^2 + 2 \cdot 4 \cdot \left(\frac{2}{3}\right)^3 + \cdots = 4 + 4 \cdot 2 \cdot \frac{2}{3} \cdot \frac{1}{1-(2/3)} = 20 \text{ feet.}$$

55. Since $0 < \ln x \le \sqrt{x}$ for all $x \ge 2$, $\sum_{k=2}^{\infty} \frac{1}{\ln k} \ge \sum_{k=2}^{\infty} \frac{1}{\sqrt{k}}$. Now, the previous exercise implies

that the series $\sum_{k=2}^{\infty} \frac{1}{\sqrt{k}}$ diverges. Therefore, the series $\sum_{k=2}^{\infty} \frac{1}{\ln k}$ also diverges.

57. The series $\sum_{k=1}^{\infty} a_k$ converges because its partial sums are bounded above (by 100) and increasing (the terms of the series are positive). Therefore, Theorem 6 implies that $\lim_{k \to \infty} a_k = 0$.

59. (a) $\sum_{k=1}^{\infty} a_k = \lim_{n \to \infty} S_n = 3$ (by the Squeeze Theorem).

    (b) Since the series converges, $\lim_{k \to \infty} a_k = 0$.

61. (a) Since $a_k \ge 0$, the partial sums of the series $\sum_{k=1}^{\infty} a_k$ form a monotonically increasing sequence. Since the series diverges, the partial sums must increase without bound.

(b) If $a_k = (-1)^k$, $S_n = -1$ when n is odd and $S_n = 0$ when n is even. Thus, $\lim\limits_{n \to \infty} S_n$ doesn't exist.

63. (a) $a_1 = \sum\limits_{j=1}^{1} \dfrac{1}{1+j} = \dfrac{1}{2}$; $a_2 = \sum\limits_{j=1}^{2} \dfrac{1}{2+j} = \dfrac{7}{12}$; $a_3 = \sum\limits_{j=1}^{4} \dfrac{1}{4+j} = \dfrac{533}{840}$.

(b) $H_8 = H_{2^3} = 1 + \sum\limits_{m=1}^{3} a_m = 1 + a_1 + a_2 + a_3 = \dfrac{761}{280}$.

(c) $a_k = \sum\limits_{j=1}^{2^{k-1}} \dfrac{1}{2^{k-1}+j} \geq \sum\limits_{j=1}^{2^{k-1}} \dfrac{1}{2^{k-1}+2^{k-1}} = \dfrac{1}{2}$.

(d) $H_{2^n} = 1 + \sum\limits_{k=1}^{n} a_k \geq 1 + \sum\limits_{k=1}^{n} \dfrac{1}{2} = 1 + n/2 \implies \lim\limits_{n \to \infty} H_{2^n} = \infty$.

§11.3 Testing for Convergence; Estimating Limits

1. (a) When $k \geq 0, k + 2^k \geq 2^k \implies a_k \leq 1/2^k = 2^{-k}$. Since $\sum_{k=0}^{\infty} 2^{-k}$ converges, the

comparison test implies that $\sum_{k=0}^{\infty} a_k$ converges.

   (b) $R_{10} = \sum_{k=11}^{\infty} a_k < \sum_{k=11}^{\infty} 2^{-k} = 2^{-11} \sum_{k=0}^{\infty} 2^{-k} = 2^{-10}$. Furthermore, since $a_k > 0$ for all
   $k \geq 11, R_{10} \geq 0$.

   (c) Since $R_{10} < 2^{-10} \approx 0.00097656$, S_n has the desired accuracy if $n \geq 10$.

   $$S_{10} = \sum_{k=0}^{10} a_k = \frac{127807216183}{75344540040} \approx 1.6963.$$

   (d) No. Since the terms of the series are all positive, the estimate in part (c) *underestimates* the limit.

3. $\sum_{k=2}^{n} a_k < \int_{1}^{n} a(x)\, dx < \sum_{k=1}^{n-1} a_k.$ [HINT: Draw pictures like those on p. 569.]

5. Draw a picture illustrating the left sum approximation L_n to the integral $\int_{1}^{n+1} a(x)\, dx$.
    Since the integrand is a decreasing function, L_n overestimates the value of the integral.

7. Draw a picture that illustrates a right Riemann sum approximation to the integral
    $\int_{n}^{\infty} a(x)\, dx$. The right sum is $\sum_{k=n+1}^{\infty} a_k$; it underestimates $\int_{n}^{\infty} a(x)\, dx$ since the integrand is

    a decreasing function. For the same reason, $a_{n+1} + \int_{n+1}^{\infty} a(x)\, dx \leq \int_{n}^{\infty} a(x)\, dx$. Finally,

    the first inequality established in this solution implies that $\sum_{k=n+2}^{\infty} a_k \leq \int_{n+1}^{\infty} a(x)\, dx$. Adding
    a_{n+1} to both sides of this inequality establishes that
    $$\sum_{k=n+1}^{\infty} a_k \leq a_{n+1} + \int_{n+1}^{\infty} a(x)\, dx.$$

9. $\int_{1}^{\infty} \frac{dx}{x^3} \leq \sum_{k=1}^{\infty} \frac{1}{k^3} \leq 1 + \int_{1}^{\infty} \frac{dx}{x^3} \implies \frac{1}{2} \leq \sum_{k=1}^{\infty} \frac{1}{k^3} \leq \frac{3}{2}.$

11. $\int_{1}^{\infty} xe^{-x}\, dx = -e^{-x}(x+1)\Big|_{1}^{\infty} = 2e^{-1} \implies \sum_{j=1}^{\infty} je^{-j}$ converges and

    $2e^{-1} \leq \sum_{j=1}^{\infty} je^{-j} \leq 3e^{-1}.$

13. (a) No. The function $a(x) = e^{\sin x}/x^2$ is *not* decreasing on the interval $[1, \infty)$ so it does not satisfy the hypotheses of the theorem.

(b) For all $k \geq 1$, $0 < \dfrac{e^{\sin k}}{k^2} \leq \dfrac{e}{k^2}$. Therefore, since $\displaystyle\sum_{k=1}^{\infty} \dfrac{e}{k^2} = e \sum_{k=1}^{\infty} \dfrac{1}{k^2}$ converges, the

comparison test implies that $\displaystyle\sum_{k=1}^{\infty} \dfrac{e^{\sin k}}{k^2}$ converges.

15. Since $\displaystyle\int_2^{\infty} \dfrac{dx}{(\ln x)^2} \geq \int_2^{\infty} \dfrac{dx}{x \ln x} = \infty$, the series $\displaystyle\sum_{n=2}^{\infty} \dfrac{1}{(\ln n)^2}$ diverges by the integral test.

17. $S_1 = \dfrac{1}{2} < \displaystyle\sum_{n=1}^{\infty} \dfrac{1}{n^2 + \sqrt{n}} < \sum_{k=1}^{\infty} \dfrac{1}{n^2} \leq 2.$

19. $\dfrac{1}{\sqrt{2}} = \dfrac{1}{\sqrt{1 + 1^2}} < \displaystyle\sum_{m=1}^{\infty} \dfrac{1}{m\sqrt{1 + m^2}} < \sum_{m=1}^{\infty} \dfrac{1}{m^2} \leq 2.$

21. Let $a_j = \dfrac{j^2}{j!}$. Since

$$\lim_{j \to \infty} \frac{a_{j+1}}{a_j} = \lim_{j \to \infty} \frac{\frac{(j+1)^2}{(j+1)!}}{\frac{j^2}{j!}} = \lim_{j \to \infty} \frac{(j+1)^2}{j^2} \cdot \frac{j!}{(j+1)!} = \lim_{j \to \infty} \frac{j+1}{j^2} = 0 < 1, \sum_{j=0}^{\infty} a_j$$

converges.

23. $\displaystyle\lim_{n \to \infty} \frac{\frac{(n+1)^2}{2^{n+1}}}{\frac{n^2}{2^n}} = \lim_{n \to \infty} \left(\frac{n+1}{n}\right)^2 \cdot \frac{1}{2} = \frac{1}{2} \implies \sum_{n=1}^{\infty} \frac{n^2}{2^n}$ converges.

25. (a) $\ln x / x$ is decreasing on $[3, \infty)$ and $\displaystyle\int_3^{\infty} \frac{\ln x}{x}\, dx$ diverges, so the integral test implies

that $\displaystyle\sum_{k=3}^{\infty} \frac{\ln k}{k}$ diverges. Therefore, since $\displaystyle\sum_{k=1}^{\infty} \frac{\ln k}{k} = \frac{1}{2} \ln 2 + \sum_{k=3}^{\infty} \frac{\ln k}{k}$, $\displaystyle\sum_{k=1}^{\infty} \frac{\ln k}{k}$ diverges.

(b) Since $1 - 1/k \leq \ln k$ for all $k \geq 1$, $1/k - 1/k^2 \leq (\ln k)/k$ for all $k \geq 1$. Therefore,

since $\displaystyle\sum_{k=1}^{\infty} \left(\frac{1}{k} - \frac{1}{k^2}\right)$ diverges, $\displaystyle\sum_{k=1}^{\infty} \frac{\ln k}{k}$ diverges.

(c) No, because $\displaystyle\lim_{k \to \infty} \frac{a_{k+1}}{a_k} = 1.$

27. (a) The ratio $\dfrac{a_{k+1}}{a_k}$ is 1 when k is odd and $1/2$ when k is even. Thus, $\displaystyle\lim_{k \to \infty} \frac{a_{k+1}}{a_k}$ doesn't exist.

(b) It converges. Let S_n denote the partial sum of the first n terms of the series. When

$$n = 2m, \ S_n = 2 - \left(\frac{1}{2}\right)^{m-1}; \text{ when } n = 2m + 1, \ S_n = 2 - 3\left(\frac{1}{2}\right)^{m+1}. \text{ It follows that}$$

$$\lim_{n\to\infty} S_n = 2.$$

29. Let $a_n = n^{-n}$. Then $\dfrac{a_{n+1}}{a_n} = \dfrac{(n+1)^{-(n+1)}}{n^{-n}} = \dfrac{n^n}{(n+1)^{(n+1)}} = \left(\dfrac{n}{n+1}\right)^n \dfrac{1}{n+1}$. Since

$$\lim_{n\to\infty}\left(\frac{n}{n+1}\right)^n = \frac{1}{e} \text{ and } \lim_{n\to\infty}\frac{1}{n+1} = 0, \ \lim_{n\to\infty}\frac{a_{n+1}}{a_n} = 0. \text{ Thus, the ratio test implies that}$$

the series $\displaystyle\sum_{n=1}^{\infty} n^{-n}$ converges.

31. Converges—by the comparison test: $\displaystyle\sum_{k=1}^{\infty}\frac{1}{k^2+3} \le \sum_{k=1}^{\infty}\frac{1}{k^2}$. (The series on the right side of the inequality is a convergent p-series.)

Since $R_N = \displaystyle\sum_{k=N+1}^{\infty}\frac{1}{k^2+3} \le \sum_{k=N+1}^{\infty}\frac{1}{k^2} \le \int_N^{\infty}\frac{dx}{x^2} = \frac{1}{N} \le 0.001$ when $N \ge 1000$,

$n \ge 1000$ implies that S_n approximates the sum of the series within 0.001.

33. Diverges—by the nth term test: $\displaystyle\lim_{k\to\infty}\frac{1}{2+\cos k}$ doesn't exist.

$$S_n = \sum_{k=0}^{n}\frac{1}{2+\cos k} \ge \sum_{k=0}^{n}\frac{1}{3} = \frac{n+1}{3}, \text{ so } S_n \ge 1000 \text{ when } n \ge 2999.$$

35. The comparison $\displaystyle\sum_{k=1}^{\infty}\frac{k}{k^6+17} < \sum_{k=1}^{\infty}\frac{k}{k^6} = \sum_{k=1}^{\infty}\frac{1}{k^5}$ shows that the original series converges.

Since $R_n \le \displaystyle\int_n^{\infty}\frac{x}{x^6+17}\,dx < \int_n^{\infty}\frac{dx}{x^5} = \frac{1}{4n^4} < 0.001$ when $n \ge 4$, the estimate

$$\sum_{k=1}^{4}\frac{k}{k^6+17} \approx 0.08524 \text{ is guaranteed to be in error by no more than } 0.001.$$

37. (a) Let $M = L + 1$ where $L = \displaystyle\sum_{k=1}^{\infty} b_k = \lim_{n\to\infty} T_n$. Since $0 \le a_k \le b_k$ for all $k \ge 1$, $S_n \le T_n$

for all $n \ge 1$. Furthermore, since $0 \le b_k$ for all $k \ge 1$, $\{\,T_n\,\}$ is an increasing sequence whose limit is $L < M$. It follows that $S_n \le T_n < M$ for all $n \ge 1$.

(b) S_n is an increasing sequence because $0 \le a_k$ for all $k \ge 1$: $S_{n+1} = S_n + a_{n+1} > S_n$.

(c) Together, parts (a) and (b) imply that $\{\,S_n\,\}$ is an increasing sequence that is bounded

above. Since $\{\,S_n\,\}$ is a sequence of partial sums of the infinite series $\displaystyle\sum_{k=1}^{\infty} a_k$,

convergence of the sequence implies convergence of the series.

39. (a) $S_{n+1} = \sum_{k=1}^{n+1} a_k = S_n + a_{n+1} \geq S_n$ since $a_k = a(k) \geq 0$ for all integers $k \geq 1$.

(b) $\int_1^n a(x)\,dx \leq \int_1^\infty a(x)\,dx$ because $a(x) \geq 0$ for all $x \geq 1$.

(c) Part (a) implies that $\{\,S_n\,\}$ is an increasing sequence. Since $S_n \leq a_1 + \int_1^n a(x)\,dx$,

part (b) implies that the sequence of partial sums is bounded above. Thus, the sequence of partial sums converges to a limit.

41. **Yes.** Let the series in question be $\sum_{k=0}^\infty a_k$ (i.e., $a_0 = 1$, $a_1 = 1/3$, $a_2 = 1/15$, $a_3 = 1/105$,

etc.). Then, $a_k \leq 3^{-k}$. Since $\sum_{k=0}^\infty 3^{-k}$ is a convergent geometric series, the series $\sum_{k=0}^\infty a_k$

converges by the comparison test.

43. $0 < \displaystyle\sum_{k=3}^\infty \frac{1}{(\ln k)^{\ln k}} = \sum_{k=3}^{1619} \frac{1}{(\ln k)^{\ln k}} + \sum_{k=1620}^\infty \frac{1}{(\ln k)^{\ln k}}$

$< \displaystyle\sum_{k=3}^{1619} \frac{1}{(\ln k)^{\ln k}} + \sum_{k=1620}^\infty \frac{1}{e^{2\ln k}}$

$< \displaystyle\sum_{k=3}^{1619} \frac{1}{(\ln k)^{\ln k}} + \sum_{k=1620}^\infty \frac{1}{k^2}.$

Since the last series on the right converges (it is a p-series), the desired result follows by the comparison test.

45. Converges. $\displaystyle\int_1^\infty \frac{\arctan x}{1+x^2}\,dx \leq \sum_{n=1}^\infty \frac{\arctan n}{1+n^2} \leq \frac{\pi}{8} + \int_1^\infty \frac{\arctan x}{1+x^2}\,dx$. Now,

$\displaystyle\int_1^\infty \frac{\arctan x}{1+x^2}\,dx = \int_{\pi/4}^{\pi/2} u\,du = \frac{3\pi^2}{32}$. Therefore, $\dfrac{3\pi^2}{32} \leq \displaystyle\sum_{n=1}^\infty \frac{\arctan n}{1+n^2} \leq \frac{\pi}{8} + \frac{3\pi^2}{32}$.

[NOTE: : Since the terms of the series are positive, any partial sum could also be used to provide a lower bound on the limit of the series. Thus, $a_1 = \pi/8$ could also be used as a lower bound.]

47. Diverges—by the integral test: $\displaystyle\int_1^\infty \frac{dx}{100+5x} = \frac{1}{5}\int_{105}^\infty \frac{du}{u} = \infty.$

49. Converges—by the comparison test: $\dfrac{1}{3} < \displaystyle\sum_{n=1}^\infty \frac{1}{n\cdot 3^n} < \sum_{n=1}^\infty \frac{1}{3^n} = \frac{1}{2}.$

[NOTE: : Since the terms of the series are positive, any partial sum could be used to provide a lower bound on the limit of the series.]

51. Converges—by the comparison test:

$$\sum_{j=0}^{\infty} \frac{j!}{(j+2)!} = \sum_{j=0}^{\infty} \frac{1}{(j+2)(j+1)}$$

$$= \frac{1}{2} + \sum_{j=1}^{\infty} \frac{1}{(j+2)(j+1)}$$

$$< \frac{1}{2} + \sum_{j=1}^{\infty} \frac{1}{j^2}$$

$$\le \frac{1}{2} + 1 + \int_{1}^{\infty} \frac{dx}{x^2} = \frac{5}{2}.$$

Thus, $\dfrac{1}{2} < \displaystyle\sum_{j=0}^{\infty} \dfrac{j!}{(j+2)!} \le \dfrac{5}{2}$.

[NOTE: : Since the terms of the series are positive, any partial sum could be used to provide a lower bound on the limit of the series.]

53. (a) The exercise statement implies that $a_2/a_1 \le r$ and $a_1 > 0$. Therefore, $a_2 \le a_1 r$.

    (b) $\dfrac{a_{k+1}}{a_k} \le r < 1$ for all $k \ge 1$ implies that $a_{k+1} \le a_k r < a_k$. From this it follows that $a_2 \le a_1 r$ and $a_3 \le a_2 r$. Multiplying both sides of the inequality $a_2 \le a_1 r$ by r produces $a_2 r \le a_1 r^2$. Therefore, $a_3 \le a_2 r \le a_1 r^2$.

        [NOTE: Multiplying both sides of the inequality $a_3 \le a_1 r^2$ by r leads to $a_3 r \le a_1 r^3$. Since $a_4 \le a_3 r$, we may conclude that $a_4 \le a_1 r^3$. Proceeding in a similar fashion, one finds that $a_{k+1} \le a_1 r^k$.]

    (c) $\displaystyle\sum_{k=1}^{\infty} a_k \le \sum_{k=0}^{\infty} a_1 r^k = \dfrac{a_1}{1-r}$. This implies that the given series converges.

    (d) $a_{n+k} \le a_{n+1} r^{k-1}$ for all $k \ge 1$ so $R_n = \displaystyle\sum_{k=n+1}^{\infty} a_k \le a_{n+1} \sum_{k=0}^{\infty} r^k = \dfrac{a_{n+1}}{1-r}$.

55. Let $a_n = (n!)^2/(2n)!$. Since $a_{n+1}/a_n = \dfrac{n+1}{2(2n+1)} \le \dfrac{1}{3}$ when $n \ge 1$,

$$R_N = \sum_{n=N+1}^{\infty} a_n \le a_{N+1} \sum_{n=0}^{\infty} \left(\frac{1}{3}\right)^n = \frac{3}{2} a_{N+1}. \text{ Thus, } R_N < 0.0005 \text{ when } N \ge 6.$$

§11.4 Absolute Convergence; Alternating Series

1. The series converges conditionally. (After the fifth term, the series has the same terms as the alternating harmonic series.)

3. $14.902 < S < 14.902 + \frac{1}{61} \approx 14.918$.

5. (a) $S_{50} \approx 0.23794$.

 (b) $|R_{50}| = \left| \displaystyle\sum_{n=51}^{\infty} \frac{\sin n}{n^3 + n^2 + n + 1 + \cos n} \right| \le \displaystyle\sum_{n=51}^{\infty} \left| \frac{\sin n}{n^3 + n^2 + n + 1 + \cos n} \right|$

 $\le \displaystyle\sum_{n=51}^{\infty} \frac{1}{n^3 + n^2 + n + 1 + \cos n} \le \displaystyle\sum_{n=51}^{\infty} \frac{1}{n^3} \le \displaystyle\int_{50}^{\infty} \frac{dx}{x^3}$.

 (The last step follows from the integral test.)

 (c) By part (b), $|R_{50}| \le \dfrac{1}{5000} = 0.0002$. Therefore, using the result in part (a),

 $S_{50} - 0.0002 \approx 0.23774 < S < S_{50} + 0.0002 \approx 0.23814$.

7. No — An example is $a_k = 1/k$. Then, $\displaystyle\sum_{k=1}^{\infty} a_k/k = \sum_{k=1}^{\infty} 1/k^2$ (a convergent series), but

 $\displaystyle\sum_{k=1}^{\infty} a_k = \sum_{k=1}^{\infty} 1/k$ (the divergent harmonic series).

9. The series converges absolutely by the alternating series test. Since

 $c_{n+1} = (n+1)^{-4} < 0.005$ when $n = 3$, $\left| S - \displaystyle\sum_{k=1}^{N} (-1)^k / k^4 \right| < 0.005$ when $N \ge 3$. Using

 $N = 3$, $S \approx -\dfrac{1231}{1296} \approx -0.94985$.

11. The series converges absolutely by the ratio test. $\left| S - \displaystyle\sum_{k=0}^{N} a_k \right| \le 0.005$ when $N \ge 2$ since

 $3^3/9! < 0.005$. Using $N = 2$, $S \approx -\dfrac{13}{8} = -1.625$.

13. **No**. Since $a_k \ge 0$ for all $k \ge 1$, $|a_k| = a_k$ for all $k \ge 1$.

15. When $p \le 0$, $\displaystyle\lim_{k \to \infty} \frac{\ln k}{k^p} = \infty$, so the series diverges by the nth term test (Theorem 6).

 When $p > 0$, the function $f(x) = \dfrac{\ln x}{x^p}$ is continuous, positive, and decreasing on $(e^{1/p}, \infty)$

 Therefore, the integral test (Theorem 8) implies that the series converges if and only if the

 improper integral $\displaystyle\int_{e^{1/p}}^{\infty} f(x)\, dx$ converges.

When $p \neq 1$, $\int \dfrac{\ln x}{x^p}\, dx = \dfrac{(1-p)\ln x - 1}{(1-p)^2 x^{p-1}}$ and, $\int \dfrac{\ln x}{x}\, dx = \frac{1}{2}(\ln x)^2$. Therefore, $\int_{e^{1/p}}^{\infty} \dfrac{\ln x}{x^p}\, dx$ diverges when $p \leq 1$ and converges when $p > 1$. It follows that the series converges only when $p > 1$.

17. The series converges absolutely when $p > 1$ since $\displaystyle\sum_{k=2}^{\infty} \dfrac{\ln k}{k^p}$ converges only when $p > 1$.

19. converges absolutely—$\displaystyle\sum_{j=1}^{\infty} \dfrac{1}{j^2}$ is a convergent p-series ($p = 2$). $\dfrac{3}{4} < \displaystyle\sum_{j=1}^{\infty} \dfrac{(-1)^{j+1}}{j^2} < 1$.

21. diverges—nth term test: $\displaystyle\lim_{n \to \infty} \dfrac{(-3)^n}{n^3}$ does not exist.

23. diverges—nth term test: $\displaystyle\lim_{k \to \infty} \dfrac{k}{2k+1} = \dfrac{1}{2} \neq 0$.

25. converges absolutely—$\displaystyle\lim_{j \to \infty} \dfrac{a_{j+1}}{a_j} = \lim_{j \to \infty} \dfrac{j+1}{(j^2 + 2j + 1)\cdots(j^2 + 1)} = 0$. The terms of the series are decreasing in absolute value for all $j \geq 1$. Thus,
$$0 = \sum_{j=0}^{1}(-1)^j a_j < \sum_{j=0}^{\infty}(-1)^j a_j < \sum_{j=0}^{2}(-1)^j a_j = \dfrac{1}{12}.$$

27. Let $b_k = a_{k+10^9}$. The alternating series test can be used to show that $\displaystyle\sum_{k=1}^{\infty}(-1)^{k+1} b_k$ converges. Since
$$\sum_{k=1}^{\infty}(-1)^{k+1} a_k = \sum_{k=1}^{10^9}(-1)^{k+1} a_k + \sum_{k=10^9+1}^{\infty}(-1)^{k+1} a_k = \sum_{k=1}^{10^9}(-1)^{k+1} a_k + \sum_{k=1}^{\infty}(-1)^{k+1} b_k,$$ the
series $\displaystyle\sum_{k=1}^{\infty}(-1)^{k+1} a_k$ converges.

29. Theorem 9 implies that $\displaystyle\sum_{j=1}^{\infty}(-1)^{j+1} b_j$ converges absolutely.

31. $a_k = (-1)^k/\sqrt{k}$.

33. (a) For $n = 1, 2, 3, \ldots$, $S_{2n+2} = S_{2n} + c_{2n+1} - c_{2n+2} \geq S_{2n}$ since $c_{2n+1} - c_{2n+2} \geq 0$.

    (b) For $n = 1, 2, 3, \ldots$, $S_{2n+1} = S_{2n-1} - c_{2n} + c_{2n+1} \leq S_{2n-1}$ since $c_{2n+1} - c_{2n} \leq 0$.

    (c) $S_{2m} = S_{2m-1} - c_{2m} \implies S_{2m} \leq S_{2m-1}$ since $c_{2m} \geq 0$.

(d) It follows from parts (a)–(c) that $S_2 \leq S_{2m} \leq S_{2m-1} \leq S_1$ for $m = 1, 2, 3, \ldots$. Thus, because the sequence of even partial sums is increasing and bounded above by S_1, it converges. Similarly, the sequence of odd partial sums converges because it is decreasing and bounded below by S_2.

(e) $\lim\limits_{m \to \infty} (S_{2m+1} - S_{2m}) = \lim\limits_{m \to \infty} c_{2m+1} = 0$. This implies that the limit of the sequence of even partial sums is the same as the limit of the sequence of odd partial sums.

(f) Since the sequence of even partial sums is increasing, $0 < S - S_{2m}$ is true. Also, since the sequence of odd partial sums is decreasing,

$S < S_{2m+1} = S_{2m} + c_{2m+1} \implies S - S_{2m} < c_{2m+1}$. Thus, $0 < S - S_{2m} < c_{2m+1}$.

Similarly, since the sequence of odd partial sums is decreasing, $0 < S_{2m+1} - S$. Also, since the sequence of even partial sums is increasing,

$S > S_{2m+2} = S_{2m+1} - c_{2m+2} \implies S - S_{2m+1} > -c_{2m+2}$. Thus,
$0 < S_{2m+1} - S < c_{2m+2}$.

§11.5 Power Series

1. $P_1(x) = x$, $P_2(x) = P_1(x) + x^2/2$, $P_4(x) = P_2(x) + x^3/3 + x^4/4$,
   $P_6(x) = P_4(x) + x^5/5 + x^6/6$, $P_8(x) = P_6(x) + x^7/7 + x^8/8$,
   $P_{10}(x) = P_8(x) + x^9/9 + x^{10}/10$.

3. $\displaystyle\lim_{j\to\infty} \left| \frac{(x/2)^{j+1}}{(x/2)^j} \right| = \lim_{j\to\infty} |x|/2 < 1$ when $|x| < 2$. Thus, the radius of convergence is $R = 2$.

5. $\displaystyle\lim_{k\to\infty} \left| \frac{\frac{x^{k+1}}{\sqrt{k+1}}}{\frac{x^k}{\sqrt{k}}} \right| = \lim_{k\to\infty} \frac{\sqrt{k}}{\sqrt{k+1}} \cdot |x| < 1$ when $|x| < 1$. Thus, the radius of convergence is
   $R = 1$.

7. $\left| \dfrac{a_{n+1}}{a_n} \right| = |x - 2| < 1 \implies R = 1$; interval of convergence is $(1, 3)$.

9. $R = 1$; interval of convergence is $[-6, -4)$.

11. The nth term test can be used to prove that the series diverges when $|x| \geq R$.

13. When $x = -R$ the series becomes $\displaystyle\sum_{k=1}^{\infty} \frac{(-1)^k}{k^2}$, which converges absolutely.

15. $\displaystyle\sum_{k=1}^{\infty} \frac{x^k}{k4^k}$.

17. $\displaystyle\sum_{k=1}^{\infty} \frac{(x-2)^k}{k^2 3^k}$.

19. $\displaystyle\sum_{k=1}^{\infty} \frac{(12 - x)^k}{k4^k}$.

21. By definition, the radius of convergence of a power series (whose base point is zero) is the largest value of R such that the series converges for all x such that $|x| < R$.

23. Let $z = x - 3$. Since $\displaystyle\sum_{k=0}^{\infty} a_k z^k$ converges only when $-2 < z \leq 2$, $\displaystyle\sum_{k=0}^{\infty} a_k(x-3)^k$ converges only when $1 < x \leq 5$.

25. (a) $R = 14$.

    (b) $b = 3$.

27. Cannot.

    This power series has an interval of convergence centered at $x = 0$. Since it converges for $x = -3$, it must converge if $-3 \le x < 3$. Since it diverges for $x = 7$, it must diverge if $x < -7$ or $x \ge 7$. The information given is insufficient to conclude anything about the convergence of the power series on the intervals $[-7, -3)$ and $[3, 7)$.

29. may (see discussion in solution to #27).

31. may (see discussion in solution to #27).

33. May. The information given implies that the power series converges on the interval $[-7, 3)$, diverges when $x \ge 7$, and diverges when $x < -11$. It does not imply anything about convergence or divergence on the intervals $[-11, -7)$ and $[3, 7)$.

35. may (see discussion in solution to #33).

37. may (see discussion in solution to #33).

39. The series $\displaystyle\sum_{n=0}^{\infty} \frac{2 \cdot 10^n}{3^n + 5}$ diverges by the ratio test—the limit of the ratio of successive terms of the series is $10/3 > 1$.

41. $\displaystyle f(1) - \sum_{n=0}^{N} \frac{2}{3^n + 5} = \sum_{n=N+1}^{\infty} \frac{2}{3^n + 5} < \sum_{n=N+1}^{\infty} \frac{2}{3^n} = \frac{1}{3^{N+1}} \sum_{n=0}^{\infty} \frac{2}{3^n} = \frac{2}{3^{N+1}} \cdot \frac{3}{2} = \frac{1}{3^N}$. Thus,

    since $3^{-5} < 0.01$, $\displaystyle \sum_{n=0}^{5} \frac{2}{3^n + 5} = \frac{367273}{447888} \approx 0.82001$ approximates $f(1)$ within 0.01.

43. The series defining h converges for all values of x.

45. The value of $h(1)$ is defined by an alternating series. Since $\left| \dfrac{(-1)^5}{5! + 5^3} \right| < 0.01$,

    $\displaystyle h(1) \approx \sum_{k=0}^{4} \frac{(-1)^k}{k! + k^3} \approx 0.58106$.

47. The domain of g is $[-9, 1]$.

49. $\displaystyle g(1) = \sum_{n=1}^{\infty} \frac{1}{n^3}$. Now, the integral test implies that $R_{10} \le \displaystyle\int_{10}^{\infty} \frac{dx}{x^3} = 0.005$ so

    $\displaystyle g(1) \approx \sum_{n=1}^{10} \frac{1}{n^3} \approx 1.19753$.

§11.6 Power Series as Functions

1. Since $\left|\dfrac{a_{k+1}}{a_k}\right| = \dfrac{|x|}{2}$, the radius of convergence is 2.

3. Since $\left|\dfrac{a_{k+1}}{a_k}\right| = \dfrac{(k+1)|x|}{2(k+2)}$, the radius of convergence is 2.

5. $f(x) = \dfrac{x^2}{1+x} = x^2 \displaystyle\sum_{k=0}^{\infty}(-1)^k x^k = \sum_{k=0}^{\infty}(-1)^k x^{k+2}$ [Substitute $u = -x$ into the power series representation of $(1-u)^{-1}$.]

7. $f(x) = (1+x)^{-2} = -\dfrac{d}{dx}\left((1+x)^{-1}\right) = -\dfrac{d}{dx}\left(\displaystyle\sum_{k=0}^{\infty}(-x)^k\right) = \sum_{k=1}^{\infty} k(-x)^{k-1}.$

9. $\arctan(2x) = \displaystyle\sum_{k=0}^{\infty}(-1)^k\dfrac{(2x)^{2k+1}}{2k+1}; \ R = \dfrac{1}{2}; \ P = 2x - \dfrac{8x^3}{3} + \dfrac{32x^5}{5} - \dfrac{128x^7}{7} + \dfrac{512x^9}{9}.$

11. $x^2\sin x = \displaystyle\sum_{k=0}^{\infty}(-1)^k\dfrac{x^{2k+3}}{(2k+1)!}; \ R = \infty; \ P = x^3 - \dfrac{x^5}{6} + \dfrac{x^7}{120} - \dfrac{x^9}{5040} + \dfrac{x^{11}}{362880}.$

13. $1/\sqrt{e} = e^{-1/2} = \displaystyle\sum_{k=0}^{\infty}\dfrac{(-1/2)^k}{k!}$. Since $1/(2^4 \cdot 4!) < 0.005$, $\displaystyle\sum_{k=0}^{3}\dfrac{(-1/2)^k}{k!} = \dfrac{29}{48} \approx 0.60417$ approximates $1/\sqrt{e}$ within 0.005.

15. $(\sin x - x)^3 = \left(\displaystyle\sum_{k=1}^{\infty}(-1)^k\dfrac{x^{2k+1}}{(2k+1)!}\right)^3 = \left(-\dfrac{x^3}{3!} + \dfrac{x^5}{5!} \mp \cdots\right)^3 = -\dfrac{x^9}{(3!)^3} + \dfrac{x^{11}}{1440} \mp \cdots.$

$(1 - \cos x)^4 = \left(\displaystyle\sum_{k=1}^{\infty}(-1)^{k+1}\dfrac{x^{2k}}{(2k)!}\right)^4 = \left(\dfrac{x^2}{2!} - \dfrac{x^4}{4!} \pm \cdots\right)^4 = \dfrac{x^8}{2^4} - \dfrac{x^{10}}{48} \pm \cdots.$

Therefore,

$$\lim_{x\to 0}\dfrac{(\sin x - x)^3}{x(1-\cos x)^4} = \lim_{x\to 0}\dfrac{-x^9/216 + x^{11}/1440 \mp \cdots}{x\cdot(x^8/16 - x^{10}/48 \pm \cdots)}$$

$$= \lim_{x\to 0}\dfrac{-1/216 + x^2/1440 \mp \cdots}{1/16 - x^2/48 \pm \cdots} = -\dfrac{2}{27}.$$

17. $\dfrac{1}{2+x} = \dfrac{1}{2}\left(\dfrac{1}{1+(x/2)}\right) = \dfrac{1}{2}\displaystyle\sum_{k=0}^{\infty}(-1)^k\left(\dfrac{x}{2}\right)^k; \ R = 2.$

19. $\sin x + \cos x = \displaystyle\sum_{k=0}^{\infty}(-1)^k\left(\dfrac{x^{2k}}{(2k)!} + \dfrac{x^{2k+1}}{(2k+1)!}\right); \ R = \infty.$

21. $(x^2 - 1)\sin x = (x^2 - 1)\sum_{k=0}^{\infty}(-1)^k \dfrac{x^{2k+1}}{(2k+1)!}$

$$= -x + \sum_{k=1}^{\infty}(-1)^{k+1}\dfrac{((2k+1)(2k)+1)x^{2k+1}}{(2k+1)!}$$

$$= \sum_{k=0}^{\infty}(-1)^{k+1}\dfrac{(4k^2+2k+1)x^{2k+1}}{(2k+1)!}; \; R = \infty.$$

23. $y = e^x = \sum_{k=0}^{\infty}\dfrac{x^k}{k!} \implies y' = \sum_{k=0}^{\infty}\dfrac{kx^{k-1}}{k!} = \sum_{k=1}^{\infty}\dfrac{x^{k-1}}{(k-1)!} = \sum_{k=0}^{\infty}\dfrac{x^k}{k!} = y.$

25. $y = e^{3x} = \sum_{k=0}^{\infty}\dfrac{(3x)^k}{k!} \implies y' = 3\sum_{k=0}^{\infty}\dfrac{k(3x)^{k-1}}{k!} = 3\sum_{k=1}^{\infty}\dfrac{(3x)^{k-1}}{(k-1)!} = 3\sum_{k=0}^{\infty}\dfrac{(3x)^k}{k!} = 3y.$

27. $y = (1-x)^{-1} = \sum_{k=0}^{\infty}x^k \implies y' = \sum_{k=0}^{\infty}kx^{k-1} = \sum_{k=0}^{\infty}(k+1)x^k = \left(\sum_{k=0}^{\infty}x^k\right)^2 = y^2.$

29. $\dfrac{\sin x}{x} = \sum_{k=0}^{\infty}(-1)^k\dfrac{x^{2k}}{(2k+1)!} = 1 - x^2/3! + x^4/5! - x^6/7! + \cdots \implies \lim_{x \to 0}\dfrac{\sin x}{x} = 1.$

31. $\dfrac{1-\cos x}{x^2} = x^{-2}\left(1 - \sum_{k=0}^{\infty}(-1)^k\dfrac{x^{2k}}{(2k)!}\right) = \sum_{k=0}^{\infty}(-1)^k\dfrac{x^{2k}}{(2k+2)!} \implies \lim_{x \to 0}\dfrac{1-\cos x}{x^2} = \dfrac{1}{2}.$

33. $\dfrac{\ln(1+x)-x}{x^2} = \sum_{k=0}^{\infty}(-1)^{k+1}\dfrac{x^k}{k+2} \implies \lim_{x \to 0}\dfrac{\ln(1+x)-x}{x^2} = -\dfrac{1}{2}.$

35. Let $f(x) = 1/(1-x) = \sum_{n=0}^{\infty}x^n$ if $|x| < 1$. Then

$$f'(x) = 1/(1-x)^2 = \sum_{n=1}^{\infty}nx^{n-1} = \dfrac{1}{x}\sum_{n=1}^{\infty}nx^n \text{ if } |x| < 1 \text{ and } x \neq 0. \text{ Therefore,}$$

$$\sum_{n=1}^{\infty}\dfrac{n}{2^n} = (1/2)f'(1/2) = 2.$$

37. (a) Integrating term by term, $\dfrac{1}{1-x} = \sum_{k=0}^{\infty}x^k \implies -\ln|1-x| = \sum_{k=1}^{\infty}\dfrac{x^k}{k}.$

   (b) The series converges on the interval $[-1, 1)$.

(c) When $x = 1/2$, part (a) implies that $-\ln(1/2) = \ln 2 = \sum\limits_{k=1}^{\infty} \dfrac{1}{k\, 2^k}$. Since the terms of

this series are all positive, the partial sums $S_N = \sum\limits_{k=1}^{N} \dfrac{1}{k\, 2^k}$ form an increasing sequence

that is bounded above by the sum of the series. Thus, $\ln 2 - S_N > 0$.

$$\ln 2 - S_N = \sum_{k=1}^{\infty} \frac{1}{k\, 2^k} - \sum_{k=1}^{N} \frac{1}{k\, 2^k} = \sum_{k=N+1}^{\infty} \frac{1}{k\, 2^k} \le \frac{1}{N+1} \sum_{k=N+1}^{\infty} \frac{1}{2^k}$$

$$= \frac{1}{(N+1)2^{N+1}} \sum_{k=0}^{\infty} \frac{1}{2^k} = \frac{1}{(N+1)2^N}.$$

39. The power series representations of each of the three functions are alternating series if $x > 0$, so the following inequalities are valid if $0 < x < 1$:
$x - x^2/2 < \ln(1 + x) < x - x^2/2 + x^3/3; \ x - x^3/3! < \sin x < x;$ and
$x^2/2! - x^4/4! < 1 - \cos x < x^2/2!.$

Since $(x - x^3/3!) - (x - x^2/2 + x^3/3) = x^2(1 - x)/2 > 0$ if $0 < x < 1$, the lower bound on $\sin x$ is greater than the upper bound on $\ln(1 + x)$, so $\ln(1 + x) < \sin x$ if $0 < x < 1$.
Also, since $(x - x^2/2) - x^2/2! = x(1 - x) > 0$ if $0 < x < 1$, the lower bound on $\ln(1 + x)$ is greater than the upper bound on $1 - \cos x$, so $1 - \cos x < \ln(1 + x)$ if $0 < x < 1$.

41. (a) $\displaystyle \int e^{-x^2}\, dx = \int \left(\sum_{k=0}^{\infty} \frac{(-x^2)^k}{k!} \right) dx = \sum_{k=0}^{\infty} \frac{(-1)^k}{(2k+1)\cdot k!} x^{2k+1}.$

(b) The approximation $\displaystyle \int_0^1 e^{-x^2}\, dx \approx \sum_{k=0}^{3} \frac{(-1)^k}{(2k+1)\cdot k!} = \frac{26}{35}$ has the desired accuracy

because $\dfrac{1}{9 \cdot 4!} < 0.005.$

43. $\displaystyle \int_0^1 \sqrt{x} \sin x\, dx = \int_0^1 \left(\sum_{k=0}^{\infty} (-1)^k \frac{x^{(4k+3)/2}}{(2k+1)!} \right) dx = \sum_{k=0}^{\infty} \frac{(-1)^k 2}{(4k+5)(2k+1)!}$

$\displaystyle \approx \sum_{k=0}^{2} \frac{(-1)^k 2}{(4k+5)(2k+1)!} = \frac{2557}{7020} \approx 0.36425.$

45. $\displaystyle e^{2x} \ln(1 + x^3) = \left(\sum_{k=0}^{\infty} \frac{(2x)^k}{k!} \right) \left(\sum_{k=1}^{\infty} (-1)^{k+1} \frac{x^{3k}}{k} \right) = x^3 + 2x^4 + 2x^5 + \frac{5}{6}x^6 + \cdots.$

47. $\displaystyle \frac{e^x}{1-x} = \left(\sum_{k=0}^{\infty} \frac{x^k}{k!} \right) \left(\sum_{k=0}^{\infty} x^k \right) = 1 + 2x + \frac{5}{2}x^2 + \frac{8}{3}x^3 + \cdots.$

49. $e^{\sin x} = \displaystyle\sum_{k=0}^{\infty} \frac{(\sin x)^k}{k!} = 1 + \sin x + \frac{\sin^2 x}{2!} + \frac{\sin^3 x}{3!} + \frac{\sin^4 x}{4!} + \cdots$

$= 1 + \left(x - x^3/3! + \cdots\right) + \frac{1}{2}\left(x - x^3/3! + \cdots\right)^2 + \frac{1}{3!}\left(x - \cdots\right)^3 + \frac{1}{4!}\left(x - \cdots\right)^4$

$= 1 + x + x^2/2 - x^4/8 + \cdots .$

51. $\displaystyle\sum_{k=1}^{\infty} kx^{k-1} = \left(\frac{1}{1-x}\right)' = \frac{1}{(1-x)^2}.$

53. $\displaystyle\sum_{k=1}^{\infty}(-1)^{k+1}x^k = 1 + \sum_{k=0}^{\infty}(-1)^{k+1}x^k = 1 - \sum_{k=0}^{\infty}(-1)^k x^k = 1 - \frac{1}{1+x} = \frac{x}{1+x}.$

§11.7 Taylor Series

1. (a) $f'(x) = \sqrt{x}e^{-x}$, $f''(x) = e^{-x}\left(\frac{1}{2\sqrt{x}} - \sqrt{x}\right)$, and $f'''(x) = e^{-x}\left(\sqrt{x} - \frac{1}{\sqrt{x}} - \frac{1}{4x^{3/2}}\right)$.

Therefore, $f(3) = 0$, $f'(3) = \sqrt{3}e^{-3}$, $f''(3) = -\frac{5\sqrt{3}}{6}e^{-3}$, and $f'''(3) = \frac{23\sqrt{3}}{36}e^{-3}$. It follows from Taylor's Theorem that

$$f(x) \approx \sqrt{3}e^{-3}(x-3) - \frac{5}{12}\sqrt{3}e^{-3}(x-3)^2 + \frac{23}{216}\sqrt{3}e^{-3}(x-3)^3.$$

(b) If $3 \leq x \leq 3.5$, $\left|f^{(4)}(x)\right| \leq 0.035$, so the approximation error is bounded by

$$\frac{0.035 \cdot (0.5)^4}{4!} \approx 9.1146 \times 10^{-5}.$$

3. The Maclaurin series representation of f is

$$f(x) = \sum_{k=0}^{\infty} \frac{(2x)^k}{k!} = \sum_{k=0}^{\infty} \frac{f(k)(0)}{k!}x^k.$$

Thus, the coefficient of x^{100} in the Maclaurin series representation of f is $2^{100}/100!$.

5. (a) $\dfrac{1}{2+x} = \dfrac{1}{2} \cdot \dfrac{1}{1+(x/2)} = \dfrac{1}{2} \cdot \dfrac{1}{1-(-x/2)} = \dfrac{1}{2}\displaystyle\sum_{k=0}^{\infty}\left(\dfrac{-x}{2}\right)^k = \displaystyle\sum_{k=0}^{\infty}\dfrac{(-1)^k x^k}{2^{k+1}}$.

(b) The coefficient of x^{259} in the Maclaurin series representation of $f(x)$ is $f^{(259)}(0)/259!$. Thus,

$$f^{(259)}(0) = -259!/2^{260}.$$

7. $K_{n+1} = e^x$ so $\left|e^x - P_n(x)\right| \leq \dfrac{e^x \cdot |x|^{n+1}}{(n+1)!} \to 0$ as $n \to \infty$.

9. (a) If $\left|f^{(n)}(x)\right| \leq n$ for all $n \geq 1$, then $K_{n+1} \leq n+1$ and

$$|f(x) - P_n(x)| \leq \frac{(n+1)|x|^{n+1}}{(n+1)!} \to 0 \text{ as } n \to \infty.$$

(b) Yes, because $\displaystyle\lim_{n\to\infty} \frac{2^{n+1}|x|^{n+1}}{(n+1)!} = 0$ for all x.

11. (a) $f(x) = x^{-1}\sin x = x^{-1}\displaystyle\sum_{k=0}^{\infty}\dfrac{(-1)^k x^{2k+1}}{(2k+1)!} = \displaystyle\sum_{k=0}^{\infty}\dfrac{(-1)^k x^{2k}}{(2k+1)!}$.

(b) The power series in part (a) converges for values of x in the interval $(-\infty, \infty)$.

(c) $f'''(x) = \displaystyle\sum_{k=2}^{\infty}\dfrac{(-1)^k(2k)(2k-1)(2k-2)x^{2k-3}}{(2k+1)!}$ for any $x \in (-\infty, \infty)$. Since $f'''(1)$ is represented by an alternating series,

$$\left|f'''(1) - \sum_{k=2}^{3}\frac{(-1)^k(2k)(2k-1)(2k-2)}{(2k+1)!}\right| < \frac{8 \cdot 7 \cdot 6}{9!} = \frac{1}{1080} < 0.005;$$

$$f'''(1) \approx \frac{37}{210} \approx 0.17619.$$

§11.8   Chapter Summary

1. $\displaystyle\lim_{k\to\infty} a_k = \infty.$

3. $\displaystyle\lim_{k\to\infty} a_k = \infty.$

5. $\displaystyle\lim_{n\to\infty} a_n = \lim_{n\to\infty} \frac{n+2}{n^3+4} = \lim_{n\to\infty} \frac{1}{3n^2} = 0.$

7. $\displaystyle\lim_{k\to\infty} a_k = 0.$

9. Converges absolutely — $\displaystyle\sum_{k=0}^{\infty} \left(\frac{1}{k!}\right)^2 \le \sum_{k=0}^{\infty} \frac{1}{k!} = e.$

11. Diverges by the nth term test: $\displaystyle\lim_{n\to\infty} \sum_{k=1}^{n} k^{-1} = \infty.$

13. Converges—by the integral test: $\displaystyle\sum_{j=1}^{\infty} j \cdot 5^{-j} \le \frac{1}{5} + \int_{1}^{\infty} x \cdot 5^{-x}\, dx = \frac{1}{5} + \frac{1 + \ln 5}{5(\ln 5)^2}.$

    NOTE: This series can also be shown to converge via the ratio test:

    $$\lim_{j\to\infty} \frac{\frac{j+1}{5^{j+1}}}{\frac{j}{5^j}} = \lim_{j\to\infty} \frac{j+1}{j} \cdot \frac{5^j}{5^{j+1}} = \lim_{j\to\infty} \frac{j+1}{5j} = \frac{1}{5} < 1.$$

    Furthermore, since $a_{j+1}/a_j = (j+1)/5j \le 2/5$ when $j \ge 1$, $\displaystyle\sum_{j=1}^{\infty} j \cdot 5^{-j} \le \sum_{j=1}^{\infty} \frac{2^{j-1}}{5^j} = \frac{1}{3}.$

15. Converges—by the integral test: $\displaystyle\int_{0}^{\infty} e^{-x^2}\, dx = \sqrt{\pi}/2.$

    $$\sum_{m=0}^{\infty} e^{-m^2} \le 1 + \int_{0}^{\infty} e^{-x^2}\, dx = 1 + \sqrt{\pi}/2.$$

17. Diverges—by the comparison test: $\displaystyle\sum_{k=1}^{\infty} \frac{k!}{(k+1)!-1} > \sum_{k=1}^{\infty} \frac{k!}{(k+1)!} = \sum_{k=1}^{\infty} \frac{1}{k+1} = \sum_{k=2}^{\infty} \frac{1}{k}.$

19. Converges—by the comparison test: $\displaystyle\sum_{k=1}^{\infty} \frac{\sqrt{k}}{k^2+1} < \sum_{k=1}^{\infty} \frac{1}{k^{3/2}} \le 1 + \int_{1}^{\infty} x^{-3/2}\, dx = 3.$

21. The series converges absolutely by the comparison test: $\displaystyle\sum_{k=0}^{\infty} \frac{1}{(k+1)\,2^k} < \sum_{k=0}^{\infty} \frac{1}{2^k} = 2.$

    $\displaystyle\left| S - \sum_{k=0}^{N} a_k \right| \le 0.005$ when $N \ge 5$ since $1/(7 \cdot 2^6) = 1/448 < 0.005.$ Using $N = 5,$

$$S \approx \frac{259}{320} = 0.809375.$$

23. converges conditionally— $\sum_{n=1}^{\infty} \frac{\cos(n\pi)}{n} = \sum_{n=1}^{\infty} \frac{(-1)^n}{n}$ which is (almost) the alternating

    harmonic series.
    $$-1 < \sum_{n=1}^{\infty} \frac{\cos(n\pi)}{n} = -\ln 2 < -\frac{1}{2}.$$

25. Diverges by the nth term test. (Since $0 < \pi - e$, $\lim_{k \to \infty} k^{\pi - e} = \infty$.)

27.
$$\sum_{j=0}^{\infty} \left(\frac{1}{2^j} + \frac{1}{3^j} \right)^2 = \sum_{j=0}^{\infty} \left(\frac{1}{2^{2j}} + \frac{2}{6^j} + \frac{1}{3^{2j}} \right) = \sum_{j=0}^{\infty} \left(\frac{1}{4^j} + \frac{2}{6^j} + \frac{1}{9^j} \right) = \frac{4}{3} + \frac{12}{5} + \frac{9}{8} = \frac{583}{120}.$$

29. Diverges by the nth term test: $\lim_{m \to \infty} \int_0^m e^{-x^2}\, dx = \sqrt{\pi}/2 \neq 0$.

31. **No**. Using the series representation of $\sin x$ and the alternating series test,
    $\sin(1/n) > 1/n - 1/6n^3 = (6n^2 - 1)/6n^3 \geq 5/6n$ for all $n \geq 1$. Thus,
    $$\sum_{n=1}^{\infty} \sin(1/n) > \frac{5}{6} \sum_{n=1}^{\infty} \frac{1}{n} = \infty.$$

33. **No**. Since $\lim_{n \to \infty} e^{-1/n} = 1 \neq 0$, the series diverges by the n-th term test.

35. (a) $\ln(n!) = \ln n + \ln(n-1) + \ln(n-2) + \cdots + \ln 2 + \ln 1$. Thus, (since $\ln 1 = 0$) $\ln(n!)$
        is a right sum approximation so $\int_1^n \ln x\, dx$. Since $\ln x$ is an increasing function on the
        interval $[1, n]$, $\ln(n!) > \int_1^n \ln x\, dx = n \ln n - n + 1$. Therefore,
        $n! > e^{n \ln n - n + 1} = n^n e^{1-n}$.

    (b) Part (a) implies that $\frac{b^N}{N!} < \left(\frac{be}{N} \right)^N \frac{1}{e}$. Thus, $\frac{b^N}{N!} < \frac{1}{2}$ when $\left(\frac{be}{N} \right)^N < \frac{e}{2}$. Therefore,
        since $\frac{e}{2} > 1$, any $N > be$ satisfies the given inequality.

37. $\left| \frac{a_{m+1}}{a_m} \right| = \frac{m^2 + 1}{(m+1)^2 + 1} \cdot |x| < 1 \implies R = 1$. The interval of convergence is $[-1, 1]$.

39. The interval of convergence is $(-1/3, 1/3)$.

41. The interval of convergence is $[-1/3, 1/3)$.

43. $(1, 5)$.

45. $[-3, 5)$.

47. $[-6, -4]$.

49. **Cannot** be true. The interval of convergence of a power series is symmetric around and includes its base point ($b = 1$ in this case).

51. **Must** be true. If the radius of convergence of the power series is 3, then the interval of convergence includes all values of x such that $|x - 1| < 3$.

53. **Cannot** be true. The interval of convergence of the power series is the solution set of the inequality $|x - 1| < 8$. Thus, the radius of convergence of the power series is 8.

55. $\dfrac{1 - \cos x}{x} = x^{-1}\left(1 - \displaystyle\sum_{k=0}^{\infty}(-1)^k \dfrac{x^{2k}}{(2k)!}\right) = \displaystyle\sum_{k=1}^{\infty}(-1)^{k+1}\dfrac{x^{2k-1}}{(2k)!} \implies \lim_{x \to 0}\dfrac{1 - \cos x}{x} = 0.$

57. $\dfrac{x - \arctan x}{x^3} = \displaystyle\sum_{k=0}^{\infty}(-1)^k \dfrac{x^{2k}}{2k + 3} \implies \lim_{x \to 0}\dfrac{x - \arctan x}{x^3} = \dfrac{1}{3}.$

59. $2^x = e^{x \ln 2} = \displaystyle\sum_{k=0}^{\infty}\dfrac{(x \ln 2)^k}{k!}; \ R = \infty.$

61. $\dfrac{5 + x}{x^2 + x - 2} = \dfrac{2}{x - 1} - \dfrac{1}{x + 2} = -\dfrac{2}{1 - x} - \dfrac{1}{2}\dfrac{1}{1 + (x/2)}$

$= -2\displaystyle\sum_{k=0}^{\infty}x^k - \dfrac{1}{2}\displaystyle\sum_{k=0}^{\infty}(-1)^k\left(\dfrac{x}{2}\right)^k$

$= -\displaystyle\sum_{k=0}^{\infty}\left(\dfrac{2^{k+2} + (-1)^k}{2^{k+1}}\right)x^k; \quad R = 1.$

63. $\dfrac{1}{1 - x} = -\dfrac{1}{1 + (x - 2)} = -\displaystyle\sum_{k=0}^{\infty}(-1)^k(x - 2)^k \implies a_k = (-1)^{k+1}.$

65. (a) **No.** The Maclaurin series representation of f is
$$f(x) = f(0) + f'(0)x + \dfrac{f''(0)}{2}x^2 + \cdots.$$ Since f is concave down at $x = 0$, the coefficient of x^2 in the Maclaurin series representation of f is negative.

   (b) **Yes.** $g''(0) = \dfrac{3}{4}\left(f'(0)\right)^2\left(g(0)\right)^5 - \dfrac{1}{2}f''(0)\left(g(0)\right)^3 > 0.$

67. Since $\displaystyle\lim_{n \to \infty}\left|\dfrac{a_{n+1}}{a_n}\right| = \lim_{n \to \infty}\dfrac{|r - n| \cdot |x|}{n + 1} < 1$ if $|x| < 1$, the binomial series converges if $|x| < 1$.

69. The binomial series for $(1 + u)^3$ terminates after a finite number of terms since $r = 3$ is an integer: $(1 + u)^3 = 1 + 3u + 3u^2 + u^3$. Therefore,
$f(x) = \left(1 + x^4\right)^3 = 1 + 3x^4 + 3x^8 + x^{12}.$

71. $g(x) = (1 + x^2)^{-3/2} \approx 1 - 3x^2/2 + 15x^4/8 - 35x^6/16.$

73. Since $\sqrt{1+u} = 1 + \dfrac{u}{2} - \dfrac{u^2}{8} + \dfrac{u^3}{16} - \dfrac{5u^4}{128} \pm \cdots$,

$\sqrt{1+x^3} = 1 + \dfrac{x^3}{2} - \dfrac{x^6}{8} + \dfrac{x^9}{16} - \dfrac{5x^{12}}{128} \pm \cdots$. Thus,

$$\int_0^{2/5} \sqrt{1+x^3}\,dx = \int_0^{2/5} \left(1 + \frac{x^3}{2} - \frac{x^6}{8} + \frac{x^9}{16} - \frac{5x^{12}}{128} \pm \cdots\right)dx$$

$$= \frac{2}{5} + \frac{1}{8}\left(\frac{2}{5}\right)^4 - \frac{1}{56}\left(\frac{2}{5}\right)^7 + \frac{1}{160}\left(\frac{2}{5}\right)^{10} - \frac{5}{1664}\left(\frac{2}{5}\right)^{13} \pm \cdots.$$

Now, since this is an alternating series (after the first term) and $2^7/(56 \cdot 5^7) < 5 \times 10^{-4}$, the value of the integral is approximated to the desired accuracy by $2/5 + 2^4/(8 \cdot 5^4) = 252/625 = 0.4032$.

75.   (a) By the binomial power series,

$$f(x) = (1+x)^r = 1 + \sum_{n=1}^{\infty} \frac{r(r-1)(r-2)\cdots(r-n+1)}{n!}x^n.$$

Thus, $g(x) = (1-x^2)^{-1/2}$ can be written as

$$1 + \sum_{n=1}^{\infty} \frac{(-1/2)(-3/2)(-5/2)\cdots(1/2-n)}{n!}(-x^2)^n$$

$$= 1 + \sum_{n=1}^{\infty} \frac{(-1)^n(-1)(-3)(-5)\cdots(1-2n)}{2^n \cdot n!}x^{2n}$$

$$= 1 + \sum_{n=1}^{\infty} \frac{1 \cdot 3 \cdot 5 \cdots (2n-1)}{2 \cdot 4 \cdot 6 \cdots (2n)}x^{2n}$$

To finish, we integrate:

$$\int\left(1 + \sum_{n=1}^{\infty} \frac{1 \cdot 3 \cdot 5 \cdots (2n-1)}{2 \cdot 4 \cdot 6 \cdots (2n)}x^{2n}\right)dx = x + \sum_{n=1}^{\infty} \frac{1 \cdot 3 \cdot 5 \cdots (2n-1)}{2 \cdot 4 \cdot 6 \cdots (2n)}\frac{x^{2n+1}}{2n+1}$$

$$= \int_0^x \frac{dt}{\sqrt{1-t^2}}$$

$$= \arcsin x.$$

(b) Let $u = \arcsin x$. Then $du = \dfrac{dx}{\sqrt{1-x^2}}$ and $\sin u = x$. Thus,

$$\int_0^1 \frac{x^{2n+1}}{\sqrt{1-x^2}}\,dx = \int_0^{\pi/2} \sin^{2n+1} u\,du$$

$$= \frac{2 \cdot 4 \cdot 6 \cdots (2n)}{3 \cdot 5 \cdot 7 \cdots (2n+1)}.$$

(c) $\displaystyle\int_0^1 \frac{\arcsin x}{\sqrt{1-x^2}}\,dx = \int_0^1 \frac{\arcsin x}{x^{2n+1}} \cdot \frac{x^{2n+1}}{\sqrt{1-x^2}}\,dx$

$\displaystyle = \int_0^1 \left(x^{-2n} + \sum_{n=1}^{\infty} \frac{1\cdot 3\cdot 5\cdots(2n-1)}{2\cdot 4\cdot 6\cdots(2n)}\frac{1}{2n+1} \right) \frac{x^{2n+1}}{\sqrt{1-x^2}}\,dx$

$\displaystyle = \int_0^1 \frac{x}{\sqrt{1-x^2}}\,dx + \left(\sum_{n=1}^{\infty} \frac{1\cdot 3\cdot 5\cdots(2n-1)}{2\cdot 4\cdot 6\cdots(2n)}\frac{1}{2n+1} \right)\int_0^1 \frac{x^{2n+1}}{\sqrt{1-x^2}}\,dx$

$\displaystyle = 1 + \sum_{k=1}^{\infty} \frac{1}{(2k+1)^2} = \sum_{k=0}^{\infty} \frac{1}{(2k+1)^2}.$

(d) Let $u = \arcsin x$. Then $du = 1/\sqrt{1-x^2}\,dx$. Thus,

$$\int_0^1 \frac{\arcsin x}{\sqrt{1-x^2}}\,dx = \int_0^{\pi/2} u\,du = \frac{\pi^2}{8}$$

(e) After writing out a few terms, it is clear that

$$\sum_{k=1}^{\infty} \frac{1}{k^2} = \sum_{k=0}^{\infty} \frac{1}{(2k+1)^2} + \sum_{k=1}^{\infty} \frac{1}{(2k)^2}.$$

Since the summation on the left hand side is a convergent p-series, we can assign to it some limit L. Then, the above equation can be rewritten as $L = \pi^2/8 + L/4$. Solving for L yields $L = \pi^2/6$. Thus, $\displaystyle\sum_{k=1}^{\infty} \frac{1}{k^2} = \frac{\pi^2}{6}$.

77. (a) Notice that $\displaystyle\lim_{m\to\infty} \sum_{k=0}^{m} \frac{(-1)^k}{k!} = e^{-1}$. Furthermore, since $\displaystyle\sum_{k=0}^{m} \frac{(-1)^k}{k!}$ is an alternating

series, $\displaystyle\left| \frac{1}{e} - \sum_{k=0}^{m} \frac{(-1)^k}{k!} \right| \le \frac{1}{(m+1)!}$. Therefore,

$$m!\left| \frac{1}{e} - \sum_{k=0}^{m} \frac{(-1)^k}{k!} \right| \le \frac{m!}{(m+1)!} = \frac{1}{m+1}.$$

(b) $m!/e = (n\cdot m!)/m = n\cdot(m-1)!$. Since this final expression is a product of integers, $m!/e$ is also an integer.

(c) Let $a_k = m!/k!$, where k is an integer such that $0 \le k \le m$. Then, $a_k = m(m-1)(m-2)\cdots(m-(m-k-1))$. Since a_k is a product of integers, it, too, is an integer. Thus, since $\displaystyle m!\sum_{k=0}^{m} \frac{(-1)^k}{k!}$ is the sum of alternatingly positive and negative integers, the summation itself is an integer.

(d) By assumption, m is a positive integer. Thus, $1/(m+1) \le 1/2$ for all m. Since N is the product of two non-negative integer values, it follows that $N = 0$.

(e) By the previous part $m! \left| \dfrac{1}{e} - \displaystyle\sum_{k=0}^{m} \dfrac{(-1)^k}{k!} \right| = 0$. Since $m! > 0$, it follows that

$$\frac{1}{e} = \sum_{k=0}^{m} \frac{(-1)^k}{k!} = \sum_{k=0}^{\infty} \frac{(-1)^k}{k!} - \sum_{k=m+1}^{\infty} \frac{(-1)^k}{k!}.$$

Thus, $\displaystyle\sum_{k=m+1}^{\infty} \frac{(-1)^k}{k!} = 0$. However, this is false, since the terms of the summation are both alternating and decreasing. Therefore, e is irrational.

79. Let $a_k = 1/k$. Then $\displaystyle\sum_{k=1}^{\infty} 2^k a_{2^k} = \sum_{k=1}^{\infty} 1$ which is obviously a divergent series. Therefore, part (c) of the previous exercise implies that $\displaystyle\sum_{k=1}^{\infty} a_k$ diverges.

81. Let $a_k = \dfrac{1}{k \, (\ln k)^p}$. The series $\displaystyle\sum_{k=2}^{\infty} a_k$ converges if and only if the series

$$\sum_{k=1}^{\infty} 2^k a_{2^k} = \sum_{k=1}^{\infty} \frac{2^k}{2^k \left(\ln 2^k \right)^p} = \sum_{k=1}^{\infty} \frac{1}{k^p (\ln 2)^p} = \frac{1}{(\ln 2)^p} \sum_{k=1}^{\infty} \frac{1}{k^p}$$

converges. By part (a), this series converges only when $p > 1$.

83. (a) Using the trigonometric identity given,

$$\frac{1}{2^k} \tan \left(\frac{x}{2^k} \right) = \frac{1}{2^k} \cot \left(\frac{x}{2^k} \right) - \frac{1}{2^{k-1}} \cot \left(\frac{x}{2^{k-1}} \right).$$

Thus, $\displaystyle\sum_{k=1}^{n} \frac{1}{2^k} \tan \left(\frac{x}{2^k} \right)$ can be written as a telescoping sum and

$$\sum_{k=1}^{n} \frac{1}{2^k} \tan \left(\frac{x}{2^k} \right) = \frac{1}{2^n} \cot \left(\frac{x}{2^n} \right) - \cot x.$$

  (b) Using l'Hôpital's Rule,

$$\lim_{n \to \infty} \sum_{k=1}^{n} \frac{1}{2^k} \tan \left(\frac{x}{2^k} \right) = \lim_{n \to \infty} \frac{1}{2^n} \cot \left(\frac{x}{2^n} \right) - \cot x = \frac{1}{x} - \cot x.$$

V   Vectors and Polar Coordinates

§V.1   Vectors and vector-valued functions

1. $u + v = (1, 2) + (2, 3) = (3, 5)$.

3. $2u - 3v = 2(1, 2) - 3(2, 3) = (-4, -5)$.

5. $|u| = \sqrt{1^2 + 2^2} = \sqrt{5}$.

7. $|3u + 2w| = |(-1, 8)| = \sqrt{(-1)^2 + 8^2} = \sqrt{65}$.

9. $u + (v + w) = u + \big((c, d) + (e, f)\big) = u + (c + e, d + f) = (a + c + e, b + d + f)$ and
    $(u + v) + w = (a + c, b + d) + w = (a + c + e, b + d + f)$. Thus,
    $u + (v + w) = (u + v) + w$.

11. (a) The diagonal vector $v + w$ has components $(3, 4)$.

   (b) The northwest-pointing diagonal vector is $w - v = (-1, 2)$.

   (c) The southeast-pointing diagonal vector is $v - w = (1, -2)$.

21. (a) $L(0) = (1, 2)$; $L(1) = (3, 5)$; $L(2) = (5, 8)$; $L(-1) = (-1, -1)$.

   (b) If the domain of L is restricted to $t \geq 0$, then the image is the ray (or half-line) that
        starts at $P(1, 2)$ and points in the same direction as the vector $(2, 3)$.

   (c) If the domain of L is restricted to $-1 \leq t \leq 1$, then the image is the line segment from
        $L(-1) = (-1, -1)$ to $L(1) = (3, 5)$.

23. The particle's velocity vector at time t is $v(t) = \left(-2 + 3t^2, 8t^3\right)$ and the particle's speed at
     time t is
     $$s(t) = |v(t)| = \sqrt{x'(t)^2 + y'(t)^2} = \sqrt{(-2 + 3t^3)^2 + (8t^3)^2}.$$
     Therefore, at time $t = 1$, the particle's speed is $\sqrt{65}$.

25. (a) The curve is the straight line joining the origin to the point $(5000, -5000)$.

   (b) Since $p(t) = \frac{t^2}{2}(1, -1)$, we have $x(t) = t^2/2$ and $y(t) = -t^2/2$. Thus $y = -x$—the
        equation of the straight line mentioned above.

   (c) For $t \geq 0$, the speed function is $s(t) = |v(t)| = |t(1, -1)| = t\sqrt{2}$. Thus the arc length
        is $\sqrt{2} \int_0^{100} t \, dt = 5000\sqrt{2} \approx 7071.07$.

27. (a) The assumption that the projectile travels under free fall conditions implies that
        $a(t) = (0, -g)$. Since the initial angle is $\pi/3$ and the initial speed is 100 meters per
        second, $v(t) = \big(100\cos(\pi/3), 100\sin(\pi/3)\big) + t(0, -g) = (50, 50\sqrt{3}) + t(0, -g) =$
        $(50, 50\sqrt{3} - gt)$. Finally, since the projectile is at $(0, 0)$ at time $t = 0$, the position
        function is $p(t) = t(50, 50\sqrt{3}) + t^2(0, -g/2)$.

(b) The projectile touches down when its y-coordinate is zero: at the time $t_0 > 0$ that is the solution of the equation $50\sqrt{3}t_0 - gt_0^2/2 = 0$. Therefore, $t_0 = 100\sqrt{3}/g \approx 17.67$.

(d) The projectile is at its maximum height when the velocity vector is horizontal (i.e., when its y-component is zero). Thus, the projectile reaches its maximum height at time $t = 50\sqrt{3}/g$. At this time, the height of the projectile is
$50^2 \cdot 3/g - 50^2 \cdot 3/2g = 50^2 \cdot 3/2g \approx 382.65$.

(e) At time $t = 50\sqrt{3}/g$, the projectile's velocity is $(50, 0)$ and it's speed is 50 meters per second.

29. (a) $\boldsymbol{a}(1) = (13, -1)$.

(b) $\boldsymbol{p}(0) = (-31/3, 0)$.

(c) $\displaystyle\int_0^2 \sqrt{25t^4 + 30t^3 - 30t^2 - 26t + 17}\, dt$.

31. (a) $\boldsymbol{v}(t) = \boldsymbol{v}(0) + (0, -t) = (1, -t)$.

(b) $\boldsymbol{p}(t) = \boldsymbol{p}(0) + (t, -t^2/2) = (t, -t^2/2)$.

(c) $\displaystyle\int_0^5 |\boldsymbol{v}(t)|\, dt = \int_0^5 \sqrt{1 + t^2}\, dt$.

33. The arc length integral is
$$\int_0^\pi \sqrt{\left(-2\sin(2t)\right)^2 + \left(2\cos(2t)\right)^2}\, dt = 2\int_0^\pi \sqrt{\sin^2(2t) + \cos^2(2t)}\, dt = 2\int_0^\pi 1\, dt = 2\pi.$$

35. The arc length integral is $\displaystyle\int_0^{2\pi} \sqrt{(3\cos t)^2 + (-\sin t)^2}\, dt = \int_0^{2\pi} \sqrt{8\cos^2 t + 1}\, dt \approx 13.365$.

37. $|\boldsymbol{r}'(t)| = \sqrt{1^2 + 3^2} = \sqrt{10}$.

39. $|\boldsymbol{r}'(t)| = \sqrt{\left(3\cos(3t)\right)^2 + \left(-3\sin(3t)\right)^2} = 3$.

41. $|\boldsymbol{r}'(t)| = \sqrt{(\cos t - t\sin t)^2 + (\sin t + t\cos t)^2} = \sqrt{1 + t^2}$.

43. In parametric form, the equation of the logarithmic spiral between $\theta = 0$ and $\theta = \pi$ is $x(t) = e^t \cos t$, $y(t) = e^t \sin t$, $0 \le t \le \pi$. Therefore, the length of the logarithmic spiral $r = e^\theta$ between $\theta = 0$ and $\theta = \pi$ is
$$\int_0^\pi \sqrt{e^{2t}(\cos t - \sin t)^2 + e^{2t}(\sin t + \cos t)^2}\, dt = \int_0^\pi \sqrt{2}e^t\, dt = \sqrt{2}\left(e^\pi - 1\right) \approx 31.312.$$

§V.2   Polar Coordinates and Polar Curves

1. $(\pi, 0)$; $(-\pi, \pi)$; $(\pi, 2\pi)$.

3. $(\sqrt{2}, \pi/4)$; $(\sqrt{2}, -7\pi/4)$; $(-\sqrt{2}, -3\pi/4)$; $(-\sqrt{2}, 5\pi/4)$.

5. $(\sqrt{5}, 1.107)$; $(\sqrt{5}, -5.176)$;
   $(-\sqrt{5}, 4.249)$.

7. $(\sqrt{17}, 1.326)$; $(\sqrt{17}, -4.957)$; $(-\sqrt{17}, 4.467)$.

9. $(\sqrt{2}, \sqrt{2})$.

11. $(\sqrt{3}/2, 1/2)$.

13. $(0.5403, 0.8415)$.

15. $(\sqrt{2}, \sqrt{2})$.

17. $x = 2$.

19. $\theta = \pi/3 \implies \tan\theta = \tan(\pi/3) = \sqrt{3} = y/x \implies y = \sqrt{3}x$.

21. $r = 3$.

23. $\tan\theta = 2$.

25. (a) The polar points $(1, 0)$, $(1, 2\pi)$, and $(-1, \pi)$ all represent the Cartesian point $(1, 0)$.
    (Use $x = r\cos\theta$ and $y = r\sin\theta$.)

    (b) The polar points $(-1, \pi/4)$, $(-1, 9\pi/4)$, and $(1, 5\pi/4)$ all represent the Cartesian
    point $(-\sqrt{2}/2, -\sqrt{2}/2)$.

    (c) $(\sqrt{2}, \pi/4 + 2k\pi)$ and $(-\sqrt{2}, \pi/4 + (2k-1)\pi)$.

27. (a)

θ	0	$\frac{\pi}{6}$	$\frac{\pi}{3}$	$\frac{\pi}{2}$	$\frac{2\pi}{3}$	$\frac{5\pi}{6}$	π	$\frac{7\pi}{6}$	$\frac{4\pi}{3}$	$\frac{3\pi}{2}$	$\frac{5\pi}{3}$	$\frac{11\pi}{6}$	2π
r	2	1.866	1.5	1	0.5	0.134	0	0.134	0.5	1	1.5	1.866	2

   (c) The cardioid is symmetric with respect to the x-axis.

   (d) The entries in the table are symmetric about $\theta = \pi$.

49. Substituting the relations $\cos\theta = x/r$ and $\sin\theta = y/r$ into the equation
    $r = a\cos\theta + b\sin\theta$ produces $r = ax/r + by/r$. Multiplying both sides of this equation by
    r leads to $r^2 = ax + by$. Therefore,
    $x^2 + y^2 = ax + by \implies (x^2 - ax) + (y^2 - by) = 0 \implies (x - a/2)^2 + (y - b/2)^2 = (a^2 + b^2)/4$.
    From this it follows that the circle has radius $\sqrt{a^2 + b^2}/2$ and center $(a/2, b/2)$.

51. (b) Since $r = \sqrt{x^2 + y^2}$ and $\cos\theta = x/r = x/\sqrt{x^2 + y^2}$,

$r = \cos^2\theta \implies \sqrt{x^2 + y^2} = x^2/(x^2 + y^2)$ or, equivalently, $(x^2 + y^2)^{3/2} = x^2$.

Squaring both sides of this equation leads to the more appealing equation

$(x^2 + y^2)^3 = x^4$.

53. The distance between the points (x_1, y_1) and (x_2, y_2) is $\sqrt{(x_2 - x_1)^2 + (y_2 - y_1)^2}$. Using $x_1 = r_1\cos\theta_1$, $x_2 = r_2\cos\theta_2$, etc.,

$$
\begin{aligned}
(x_2 - x_1)^2 + (y_2 - y_1)^2 &= (r_2\cos\theta_2 - r_1\cos\theta_1)^2 + (r_2\sin\theta_2 - r_1\sin\theta_1)^2 \\
&= r_2^2\cos^2\theta_2 + r_1^2\cos^2\theta_1 - 2r_1r_2\cos\theta_1\cos\theta_2 \\
&\quad + r_2^2\sin^2\theta_2 + r_1^2\sin^2\theta_1 - 2r_1r_2\sin\theta_1\sin\theta_2 \\
&= r_1^2 + r_2^2 - 2r_1r_2(\cos\theta_1\cos\theta_2 + \sin\theta_1\sin\theta_2) \\
&= r_1^2 + r_2^2 - 2r_1r_2\cos(\theta_1 - \theta_2).
\end{aligned}
$$

Thus, the distance between the points (x_1, y_1) and (x_2, y_2) is $\sqrt{r_1^2 + r_2^2 - 2r_1r_2\cos(\theta_1 - \theta_2)}$.

§V.3   Calculus in Polar Coordinates

1. The results are straightforward applications of the product rule: $(f \cdot g)' = f' \cdot g + f \cdot g'$.

3. (b) $\dfrac{dy}{dx} = \dfrac{\sin\theta + \theta\cos\theta}{\cos\theta - \theta\sin\theta}$. Thus, the spiral has a horizontal tangent line wherever $\theta = -\tan\theta$.

   (c) The formula for dy/dx in part (a) implies that the spiral has a vertical tangent line wherever $\theta = \cot\theta$.

   (d) The polar point $(1, 1)$ is the point $(\cos 1, \sin 1)$ in Cartesian coordinates. The slope of the tangent line at the polar point $(1, 1)$ is
   $m = (\sin 1 + \cos 1)/(\cos 1 - \sin 1) \approx -4.588$. Thus, the equation of the desired tangent line is $y = m(x - \cos 1) + \sin 1 \approx -4.588(x - 0.540) + 0.841$.

5. area $= \dfrac{1}{2}\displaystyle\int_0^\pi 1\, d\theta = \pi/2$.

7. area $= \dfrac{1}{2}\displaystyle\int_0^\pi a^2\, d\theta = \dfrac{\pi a^2}{2}$.

9. area $= \displaystyle\int_0^{\pi/2} f(\theta)^2\, d\theta + \int_{11\pi/6}^{2\pi} f(\theta)^2\, d\theta = \dfrac{1}{2}\int_{-\pi/6}^{7\pi/6} f(\theta)^2\, d\theta = 2\pi + \dfrac{3\sqrt{3}}{2}$.

11. area $= \dfrac{1}{2}\displaystyle\int_{-\pi/2}^{\pi/2} f(\theta)^2\, d\theta = \dfrac{1}{2}\int_{-\pi/2}^{\pi/2}(1 - \cos\theta)^2\, d\theta = \int_0^{\pi/2}(1 - \cos\theta)^2\, d\theta$

   $= \left(\dfrac{3\theta}{2} - 2\sin\theta + \dfrac{1}{2}\cos\theta\sin\theta\right)\Big]_0^{\pi/2} = 3\pi/4 - 2$.

13. area $= \dfrac{1}{2}\displaystyle\int_0^{\pi/2}\big(f(\theta)^2 - g(\theta)^2\big)\, d\theta + \dfrac{1}{2}\int_\pi^{2\pi} f(\theta)^2\, d\theta$

   $= \dfrac{1}{2}\displaystyle\int_0^{\pi/2}\big((1 + \cos\theta)^2 - \sin^2\theta\big)\, d\theta + \dfrac{1}{2}\int_\pi^{2\pi}(1 + \cos\theta)^2\, d\theta$

   $= \left(\dfrac{\theta}{2} + \sin\theta + \dfrac{\cos\theta\sin\theta}{2}\right)\Big]_0^{\pi/2} + \left(\dfrac{3\theta}{4} + \sin\theta + \dfrac{\cos\theta\sin\theta}{4}\right)\Big]_\pi^{2\pi}$

   $= (1 + \pi/4) + 3\pi/4 = 1 + \pi$.

15. area $= m/2$.

   area $= \dfrac{1}{2}\displaystyle\int_0^{\arctan m} \sec^2\theta\, d\theta = \dfrac{1}{2}\tan\theta\,\Big]_0^{\arctan m} = \dfrac{m}{2}$.

17. (a) One leaf lies between $\theta = -\pi/2n$ and $\theta = \pi/2n$. The area of this leaf is
   $\dfrac{1}{2}\displaystyle\int_{-\pi/2n}^{\pi/2n} \cos^2(n\theta)\, d\theta = \pi/4n$.

(b) The area of all n leaves is $\pi/4$ (i.e., one-fourth of the area of the circle $r = 1$).

19. area $= \left(e^{4\pi} - 1\right)/4$.

21. area $= \dfrac{1}{2} \displaystyle\int_{-\pi/3}^{\pi/3} d\theta - \dfrac{1}{2} \displaystyle\int_{-\pi/3}^{\pi/3} \left(\tfrac{1}{2}\sec\theta\right)^2 d\theta = \dfrac{\pi}{3} - \dfrac{\sqrt{3}}{4}$.

23. $x = 2\cos t,\ y = 2\sin t,\ 0 \le t \le 2\pi$. The graph is a circle of radius 2 with center at the origin.

25. $x = 1,\ y = \tan t,\ -\pi/4 \le t \le \pi/4$. The graph is the vertical line segment from $(1, -1)$ to $(1, 1)$.

27. $x = t\cos t,\ y = t\sin t,\ 0 \le t \le 2\pi$. The graph is one loop of a spiral.

29. $x = \cos(2t)\cos t,\ y = \cos(2t)\sin t,\ 0 \le t \le 2\pi$. The graph is the 4-leaf rose shown in Example 3.

31. $\left(\dfrac{dy}{d\theta}\right)^2 + \left(\dfrac{dx}{d\theta}\right)^2 = \left(f'(\theta)\sin\theta + f(\theta)\cos\theta\right)^2 + \left(f'(\theta)\cos\theta - f(\theta)\sin\theta\right)^2$

$$= \left(f'(\theta)^2\sin^2\theta + 2f(\theta)f'(\theta)\sin\theta\cos\theta + f(\theta)^2\cos^2\theta\right)$$
$$+ \left(f'(\theta)^2\cos^2\theta - 2f(\theta)f'(\theta)\sin\theta\cos\theta + f(\theta)^2\sin^2\theta\right)$$
$$= f'(\theta)^2 + f(\theta)^2 > 0,$$

which implies that $dx/d\theta$ and $dy/d\theta$ are not both simultaneously zero.

M   *Multivariable Calculus: A First Look*

§M.1   Three-Dimensional Space

1. (a) The surface is unrestricted in the x-direction.

    (b) The surface is unrestricted in the y-direction.

3. Two planes, crossing diagonally along the x-axis.

5. A parabolic "tent," unrestricted in the x-direction.

7. $x^2 + y^2 + z^2 = 4$.

9. $x^2 + z^2 = 1$.

11. $z = \sin x$ and $z = \cos x$ are two possibilities.

13. Completing the square in y and z gives
$x^2 + y^2 - 6y + z^2 - 4z = 0 \iff x^2 + (y-3)^2 + (z-2)^2 = 13$. Thus we have a sphere of radius $\sqrt{13}$, centered at $(0, 3, 2)$.

15. (b) According to the Pythagorean rule, $|OQ|^2 = |OP|^2 + |PQ|^2 = 1 + 4 = 5$ so the length of OQ is $\sqrt{5}$.

    (c) According to the Pythagorean rule, $|OR|^2 = |OQ|^2 + |QR|^2 = 5 + 9 = 14$ so the length of QR is $\sqrt{14}$.

    (d) Using the distance formula, $|OQ| = \sqrt{(1-0)^2 + (2-0)^2} = \sqrt{5}$ and
    $|OR| = \sqrt{(1-0^2 + (2-0)^2 + (3-0)^2} = \sqrt{14}$.

17. $d(P, Q) = \sqrt{(a-x)^2 + (b-y)^2 + (c-z)^2}$
$= \sqrt{(x-a)^2 + (y-b)^2 + (z-c)^2} = d(Q, P)$.

19. $(1, 1, 1)$, $(1, 1, -1)$, $(1, -1, 1)$, $(1, -1, -1)$, $(-1, 1, 1)$, $(-1, 1, -1)$, $(-1, -1, 1)$, $(-1, -1, -1)$.

21. The line $Ax + By = C$ has slope $-A/B$, so lines with $B = 0$ have undefined slope.

23. Setting $y = 0$ in $Ax + By = C$ gives $Ax = C$, or $x = C/A$. Thus, lines with $A = 0$ have no x-intercept.

25. Setting $y = 0$ and $z = 0$ in $Ax + By + Cz = D$ gives $Ax = D$, so the x-intercept is $x = D/A$. Thus any plane with $A = 0$ and $D \neq 0$ (such as $y + z = 1$) has no x-intercept.

27. (a) Setting $y = 0$ in $x + 2y + 3z = 3$ gives the trace $x + 3z = 3$; this line intercepts the x- and z-axes at $x = 3$ and $z = 1$, respectively.

(b) Setting $x = 0$ in $x + 2y + 3z = 3$ gives the trace $2y + 3z = 3$; this line intercepts the y- and z-axes at $y = 3/2$ and $z = 1$, respectively.

(c) Setting $x = 1$ in $x + 2y + 3z = 3$ gives the trace $1 + 2y + 3z = 3$, or $2y + 3z = 2$.

29. The line $1x + 0y = 3$ intercepts the x-axis but not the y-axis.

31. The plane $x + 2y = 1$ in xyz-space intercepts the x-axis at $x = 1$ and the y-axis at $y = 1/2$.

33. The midpoint has coordinates $((x_1 + x_2)/2, (y_1 + y_2)/2, (z_1 + z_2)/2)$; one shows that these coordinates satisfy the equation of the given plane.

§M.2 Functions of Several Variables

1. $g(x, y) = x^2 + y^2$ has domain $\mathbb{R}^2$; the range is $[0, \infty)$.

3. $j(x, y) = 1/(x^2 + y^2)$ has domain all of $\mathbb{R}^2$ except for the origin; the range is $(0, \infty)$.

5. The domain of $m(x, y) = \sqrt{1 - x^2 - y^2}$ is all points inside and on the unit circle in $\mathbb{R}^2$; the range is $[0, 1]$.

7. (a) In the rectangle All level curves of f are circles centered at the origin, except the curve $z = 0$, which is just the origin.

   (b) All level curves of g are circles centered at the origin, except for except the curve $z = 1$, which is just the origin.

   (c) The curves themselves are identical, but their labels are different.

   (d) Both graphs are paraboloids, opening upward. The g-graph is one unit higher.

9. (a) Each level curve is a vertical line (since the value of f depends only on x).

   (b) Each level curve is a horizontal line (since the value of g depends only on y).

   (c) In both parts, the level curves are lines. However, the level curves of f are vertical lines while the level curves of g are horizontal lines.

   (d) The graph of g can be obtained from the graph of f by rotating it by $\pi/2$ radians.

   (e) The contour map of a cylinder is a horizontal line or a vertical line (depending on which variable is unrestricted).

11. Level curves of T, in weather language, are curves along which the temperature remains constant. As a rule, such "isotherms" run east-and-west.

13. Near the coldest (or warmest) spot in the country, the level curves should be closed curves, shrinking down to the cold spot.

15. (a) $f(x, y) = |x - 1|$

   (b) The graph of f is a vee-shaped trough, parallel to the y-axis.

   (c) The level curve of f through $(3, 4)$ is the straight line $x = 3$.

   (d) All level curves of f are straight lines parallel to the y-axis.

§M.3 Partial Derivatives

1. $f_x(x, y) = 2x$; $f_y(x, y) = -2y$.

3. $f_x(x, y) = \dfrac{2x}{y^2}$; $f_y(x, y) = \dfrac{-2x^2}{y^3}$.

5. $f_x(x, y) = -\sin(x)\cos(y)$; $f_y(x, y) = -\cos(x)\sin(y)$.

7. $f_x(x, y, z) = y^2 z^3$; $f_y(x, y, z) = 2xyz^3$; $f_z(x, y, z) = 3xy^2 z^2$.

9. (a) Because $f(3) = 9$ and $f'(3) = 6$, it follows that the linear approximation function has equation $L(x) = 9 + 6(x - 3)$.

   (b) The graphs of f and L are close together near $x = 3$; in particular, the graph of f looks similar to a straight line.

   (c) The inequality $|f(x) - L(x)| < 0.01$ holds for x in the interval $2.9 < x < 3.1$. This can be found either graphically, by zooming, or algebraically. The latter approach can be done as follows. Notice first that

   $$|f(x) - L(x)| = |x^2 - (9 + 6(x - 3))| = |x^2 - 6x + 9| = |x - 3|^2.$$

   Thus

   $$|f(x) - L(x)| < 0.01 \iff |x - 3|^2 < 0.01 \iff |x - 3| < 0.1.$$

   This is equivalent to saying that $2.9 < x < 3.1$.

11. (a) The fact that $f'(3) = 6$ means that the graph of f looks like a line of slope 6 near $x = 3$. (The graph bears this out.)

   (b) Both of the secant lines, from $x = 3$ to $x = 3.5$ and from $x = 3$ to $x = 3.1$, have slopes near 6.

   (c) The average rates of change $\Delta y / \Delta x$ of f over the intervals $[3, 3.5]$ and $[3, 3.1]$ are 6.5 and 6.1, respectively.

   (d) For $f(x) = x^2$, $\displaystyle\lim_{h \to 0} \frac{f(3 + h) - f(3)}{h} = 6$. The answer is the derivative of f at $x = 3$.

13. $g_x(1, 1) \approx (g(1.01, 1) - g(1, 1))/0.01 = -2.01$. Similarly, $g_y(1, 1) \approx (g(1, 1.01) - g(1, 1))/0.01 = 3.01$.

15. (a) Since $L(x, y) = y$, $L_x(x, y) = 0$ and $L_y(x, y) = 1$. It follows that $L(0, 0) = 0$, $L_x(0, 0) = 0$, and $L_y(0, 0) = 1$.

   (b)

y ＼ x	−0.02	−0.01	0.00	0.01	0.02
0.02	0.02	0.02	0.02	0.02	0.02
0.01	0.01	0.01	0.01	0.01	0.01
0.00	0.00	0.00	0.00	0.00	0.00
−0.01	−0.01	−0.01	−0.01	−0.01	−0.01
−0.02	−0.02	−0.02	−0.02	−0.02	−0.02

17. (a) The graph is a cylinder in the x-direction.

(b) $f_x(x, y) = 0$; $f_y(x, y) = \cos y$. The fact that $f_x = 0$ reflects the fact that f is constant in x.

(c) The linear approximation function L for f at the point $(0, 0)$ is $L(x, y) = y + 2$. Like f itself, L is independent of x.

19. (a) All the level curves are vertical lines.

(b) >From the diagram, $f_x(0, 0) \approx 1$ and $f_y(0, 0) = 0$.

(c) >From the diagram, $f_x(\pi/2, 0) \approx 0$ and $f_y(0, 0) = 0$.

(d) All contour lines are vertical; this means that $f(x, y)$ is constant along vertical lines, which implies that $f_y(x, y) = 0$.

(e) The fact that $f_x(x, y)$ is independent of y means that the rate of increase of f in the x-direction is the same for all y. The contour map of f reflects this fact in that contour lines are all vertical.

21. (a) All the level curves are straight lines; all have the same slope, and they're equally spaced.

(b) The contour map shows that $f_x(0, 0) = 2$ and $f_y(0, 0) = -3$; one sees this by measuring how fast f increases in the x- and y-directions.

(c) The contour map of f reflects the fact that both f_x and f_y are constant functions in the sense that all contour lines are straight and equally spaced.

(d) The contour map shows that $f_x(x, y)$ is positive and $f_y(x, y)$ is negative in that moving to the right (in the x-direction) corresponds to increasing z, while moving upward (in the y-direction) corresponds to decreasing z.

23. (a) The partial derivatives of $f(x, y) = xy$ are $f_x(x, y) = y$ and $f_y(x, y) = x$. Thus, at the base point $(x_0, y_0) = (2, 1)$, $f(2, 1) = 2$, $f_x(2, 1) = 1$, and $f_y(2, 1) = 2$. Therefore $L(x, y) = 1(x - 2) + 2(y - 1) + 2 = x + 2y - 2$.

(b) Level curves of $f(x, y) = xy$ are hyperbolas, of the form $xy = k$. Level curves of $L(x, y) = x + 2y - 2$ are straight lines, of the form $x + 2y - 2 = k$.

(c) The level curves of f and L look similar near $(2, 1)$. (E.g., all the level curves of L are lines, with slope $-1/2$. Similarly, the hyperbola $xy = 2$ (a level curve of f) has slope $-1/2$ at the point $(2, 1)$.

(d) In a small window around $(2, 1)$, the level curves of f and L appear almost identical—this is because L is the linear approximation to f at $(2, 1)$.

25. (a) From the contour map of f we estimate $f_x(1, 2) = 2$ and $f_y(1, 2) = 4$.

(b) Symbolic differentiation of $f(x, y) = x^2 + y^2$ gives $f_x = 2x$ and $f_y = 2y$, so $f_x(1, 2) = 2$ and $f_y(1, 2) = 4$.

(c) It follows from the above and from the fact that $f(1, 2) = 5$ that $L(x, y) = 2(x - 1) + 4(y - 2) + 5 = 2x + 4y - 5$.

(d) Level curves $L(x, y) = k$ are all straight lines, with slope $-1/2$. Level curves $f(x, y) = k$ are all circles, centered at the origin. At the point $(1, 2)$, the level curves for f and L are tangent to each other.

27. If $f(x, y) = x^2 + y^2$ and $(x_0, y_0) = (2, 1)$, then $L(x, y) = -5 + 4x + 2y$.

29. If $f(x, y) = \sin(x) + \sin(y)$ and $(x_0, y_0) = (0, 0)$, then $L(x, y) = x + y$.

31. Suppose that f is independent of x. Then, $f_x(x, y) = 0$ at all points (x, y). Therefore, $L(x, y) = f(x_0, y_0) + f_x(x_0, y_0)(x - x_0) + f_y(x_0, y_0)(y - y_0) = f(x_0, y_0) + f_y(x_0, y_0)(y - y_0)$ which is also a function that is independent of x.

33. (a) The information given on partial derivatives implies that
$$L(x, y) = 0.6(x - 3) + 0.8(y - 4) + 5 = 0.6x + 0.8y.$$

    (b) Use the formula above: $L(2.9, 4.1) = 0.6(-0.1) + 0.8(0.1) + 5 = 5.02$;
$L(4, 5) = 0.6(1) + 0.8(1) + 5 = 6.4$.

    (c) The function g can not be linear. If g were linear it would agree everywhere with L—but we saw above that $g(4, 5) = \sqrt{41} \neq 6.4 = L(4, 5)$.

35. (a) Graphs show a crease along the y-axis; this suggests that f_x may not exist there. f_y exists everywhere.

    (b) Since $f(x, 0) = 0$, it follows that $f_x(x, 0) = 0$.

    (c) Since $f(x, 1) = |x|$, it follows that $f_x(x, 1)$ doesn't exist.

§M.4 Optimization and Partial Derivatives: A First Look

1. (a) The ant's altitude remains constant.

    (b) The ant rises all the way; it's highest at $(0.5, 1)$. At that point, its altitude is 0.5.

3. (a) The function f has stationary points $(1, \pi/2)$ and $(1, -\pi/2)$. The first is a local minimum, the second a local maximum, apparently.

    (b) The partial derivatives are $f_x(x, y) = (2x - 2)\sin(y)$; $f_y(x, y) = (x^2 - 2x)\cos(y)$. It follows that the stationary points are at $(0, 0)$, $(2, 0)$, $(1, \pi/2)$ and $(1, -\pi/2)$.

    (c) The functions $f(-3, y) = -15\sin(y)$ has minimum value -15 at $y = -\pi/2$ and maximum value 15 at $y = \pi/2$.

5. There's a saddle at $(0, 0)$.

7. There's a saddle at $(1, 2)$.

9. (a) A plane can have a stationary points only if the plane is horizontal. In this case, every point is a stationary point.

    (b) $L_x = a$ and $L_y = b$, so there are stationary points only if $a = b = 0$. (c can have any value.) In this case, every point is stationary.

11. $g(x, y) = y^2$ works; so do many others.

13. $k(x, y) = (x - 3)^2 + (y - 4)^2$ works; so do many others.

§M.5 Multiple Integrals and Approximating Sums

1. Doing this by hand gives $\frac{1}{4} f\left(\frac{1}{8}\right) + \frac{1}{4} f\left(\frac{3}{8}\right) + \frac{1}{4} f\left(\frac{5}{8}\right) + \frac{1}{4} f\left(\frac{7}{8}\right)$. Using $f(x) = x^2$, and working this out as a fraction, gives $21/64$.

   Maple's value for M_{100}, the midpoint sum with 100 subdivisions, is 0.333325.

3. The triple midpoint sum has eight terms, each of the form $8abc$, where (a, b, c) is the midpoint of one of the 8 subcubes. The total sum, evaluated numerically, is 512.

   Maple's value with $n = 2$, i.e., the triple midpoint sum with 8 subdivisions is 512. With $n = 4$, *Maple* still gives 512—this happens to be the exact answer.

5. (a) Either the level curves or the formula can be used to estimate values of f at the midpoints; they lead to the estimate $I \approx M_{16} = 168$ (or something near that).

   (b) *Maple* calculates the double midpoint sum as 168.

   (c) Since the surface rises faster and faster as away from the origin, the midpoint sum somewhat underestimates the integral.

§M.6   Calculating Integrals by Iteration

1. If $R = [0, 1] \times [0, 1]$, then $\iint_R \sin(x)\sin(y)\,dA = \left(\cos(1)\right)^2 - 2\cos(1) + 1 \approx 0.2113$.

3. If $R = [0, 4] \times [0, 4]$, then $\iint_R (x^2 + y^2)\,dA = \dfrac{512}{3} \approx 170.6667$.

5. If $V = [0, 1] \times [0, 2] \times [0, 3]$, then $\iiint_R y\,dV = 6$.

7. $\iint_R x\,dA = \displaystyle\int_{x=0}^{x=1}\int_{y=x^2}^{y=\sqrt{x}} x\,dy\,dx = \dfrac{3}{20} = 0.15$.

9. $\iint_R (x+y)\,dA = \displaystyle\int_{y=0}^{y=1}\int_{x=y}^{x=\sqrt{y}} (x+y)\,dx\,dy = \dfrac{3}{20} = 0.15$.

11. $\iint_R 1\,dA = \displaystyle\int_{y=0}^{y=1}\int_{x=0}^{x=\sqrt{1-y^2}} 1\,dx\,dy = \dfrac{\pi}{4} \approx 0.7854$.

13. (a) Integrating first in y, then in x, $\iint_R x\,dA = \displaystyle\int_{x=0}^{x=1}\int_{y=0}^{y=e^x} x\,dy\,dx = 1$.

(b) Integrating first in x, then in y, requires that the y-interval be broken up into two parts:

$$\iint_R x\,dA = \int_{y=0}^{y=1}\int_{x=0}^{x=1} x\,dx\,dy + \int_{y=1}^{y=e}\int_{x=\ln y}^{x=1} x\,dx\,dy = 1/2 + 1/2 = 1.$$

15. (a) Since R is the plane region bounded by the curves $x = g(y)$, $x = 0$, $y = c$, and $y = d$, the area of R is $\displaystyle\int_c^d g(y)\,dy$.

(b) $I = \iint_R 1\,dA = \displaystyle\int_c^d\int_0^{g(y)} 1\,dx\,dy = \int_c^d x \Big]_0^{g(y)}\,dy = \int_c^d g(y)\,dy$.

§M.7   Double Integrals in Polar Coordinates

1. (a) The area of a wedge with outer radius r_2 and making angle $\Delta\theta$ at the origin is $\dfrac{r_2{}^2}{2}\Delta\theta$.
        A polar rectangle is the difference of two such wedges; the result follows with a little computation.

    (b) If we set $a = r$, $b = r + \Delta r$, and $\beta - \alpha = \Delta\theta$ in part (a), then the result follows.

3. (a) The solid lies under a surface with parabolic cross sections.

    (b) $I_1 = 2\pi/3$, no matter how you slice it.

5. The triangle has area $1/2$; note that it's bounded by the polar curves $\theta = 0$, $\theta = \pi/4$, and $r = \sec\theta$.

7. The integral reduces to $\displaystyle\int_0^\pi (1 + \sin\theta)\, d\theta = \pi + 2$.

9. The surface lies above the unit circle in the xy-plane; the surface is defined by the function $g(r, \theta) = 1 - r$. Thus, in polar form, the integral is $\displaystyle\int_0^{2\pi}\int_0^1 (1 - r)r\, dr\, d\theta = \dfrac{\pi}{3}$.